NUCLEAR KNOWLEDGE MANAGEMENT AND HUMAN RESOURCES DEVELOPMENT: CHALLENGES AND OPPORTUNITIES

The following States are Members of the International Atomic Energy Agency:

AFGHANISTAN
ALBANIA
ALGERIA
ANGOLA
ANTIGUA AND BARBUDA
ARGENTINA
ARMENIA
AUSTRALIA
AUSTRIA
AZERBAIJAN
BAHAMAS, THE
BAHRAIN
BANGLADESH
BARBADOS
BELARUS
BELGIUM
BELIZE
BENIN
BOLIVIA, PLURINATIONAL
  STATE OF
BOSNIA AND HERZEGOVINA
BOTSWANA
BRAZIL
BRUNEI DARUSSALAM
BULGARIA
BURKINA FASO
BURUNDI
CABO VERDE
CAMBODIA
CAMEROON
CANADA
CENTRAL AFRICAN
  REPUBLIC
CHAD
CHILE
CHINA
COLOMBIA
COMOROS
CONGO
COOK ISLANDS
COSTA RICA
CÔTE D'IVOIRE
CROATIA
CUBA
CYPRUS
CZECH REPUBLIC
DEMOCRATIC REPUBLIC
  OF THE CONGO
DENMARK
DJIBOUTI
DOMINICA
DOMINICAN REPUBLIC
ECUADOR
EGYPT
EL SALVADOR
ERITREA
ESTONIA
ESWATINI
ETHIOPIA
FIJI
FINLAND
FRANCE
GABON
GAMBIA, THE

GEORGIA
GERMANY
GHANA
GREECE
GRENADA
GUATEMALA
GUINEA
GUYANA
HAITI
HOLY SEE
HONDURAS
HUNGARY
ICELAND
INDIA
INDONESIA
IRAN, ISLAMIC REPUBLIC OF
IRAQ
IRELAND
ISRAEL
ITALY
JAMAICA
JAPAN
JORDAN
KAZAKHSTAN
KENYA
KOREA, REPUBLIC OF
KUWAIT
KYRGYZSTAN
LAO PEOPLE'S DEMOCRATIC
  REPUBLIC
LATVIA
LEBANON
LESOTHO
LIBERIA
LIBYA
LIECHTENSTEIN
LITHUANIA
LUXEMBOURG
MADAGASCAR
MALAWI
MALAYSIA
MALDIVES
MALI
MALTA
MARSHALL ISLANDS
MAURITANIA
MAURITIUS
MEXICO
MONACO
MONGOLIA
MONTENEGRO
MOROCCO
MOZAMBIQUE
MYANMAR
NAMIBIA
NEPAL
NETHERLANDS,
  KINGDOM OF THE
NEW ZEALAND
NICARAGUA
NIGER
NIGERIA
NORTH MACEDONIA
NORWAY

OMAN
PAKISTAN
PALAU
PANAMA
PAPUA NEW GUINEA
PARAGUAY
PERU
PHILIPPINES
POLAND
PORTUGAL
QATAR
REPUBLIC OF MOLDOVA
ROMANIA
RUSSIAN FEDERATION
RWANDA
SAINT KITTS AND NEVIS
SAINT LUCIA
SAINT VINCENT AND
  THE GRENADINES
SAMOA
SAN MARINO
SAUDI ARABIA
SENEGAL
SERBIA
SEYCHELLES
SIERRA LEONE
SINGAPORE
SLOVAKIA
SLOVENIA
SOMALIA
SOUTH AFRICA
SPAIN
SRI LANKA
SUDAN
SWEDEN
SWITZERLAND
SYRIAN ARAB REPUBLIC
TAJIKISTAN
THAILAND
TOGO
TONGA
TRINIDAD AND TOBAGO
TUNISIA
TÜRKİYE
TURKMENISTAN
UGANDA
UKRAINE
UNITED ARAB EMIRATES
UNITED KINGDOM OF
  GREAT BRITAIN AND
  NORTHERN IRELAND
UNITED REPUBLIC OF TANZANIA
UNITED STATES OF AMERICA
URUGUAY
UZBEKISTAN
VANUATU
VENEZUELA, BOLIVARIAN
  REPUBLIC OF
VIET NAM
YEMEN
ZAMBIA
ZIMBABWE

The Agency's Statute was approved on 23 October 1956 by the Conference on the Statute of the IAEA held at United Nations Headquarters, New York; it entered into force on 29 July 1957. The Headquarters of the Agency are situated in Vienna. Its principal objective is "to accelerate and enlarge the contribution of atomic energy to peace, health and prosperity throughout the world".

PROCEEDINGS SERIES

NUCLEAR KNOWLEDGE MANAGEMENT AND HUMAN RESOURCES DEVELOPMENT: CHALLENGES AND OPPORTUNITIES

PROCEEDINGS OF AN INTERNATIONAL CONFERENCE
ORGANIZED BY THE
INTERNATIONAL ATOMIC ENERGY AGENCY
AND HELD IN VIENNA, 1–5 JULY 2024

INTERNATIONAL ATOMIC ENERGY AGENCY
VIENNA, 2026

IAEA Library Cataloguing in Publication Data

Names: International Atomic Energy Agency.
Title: Nuclear knowledge management and human resources development : challenges and opportunities / International Atomic Energy Agency.
Description: Vienna : International Atomic Energy Agency, 2026. | Series: Proceedings series (International Atomic Energy Agency), ISSN 0074–1884 | Includes bibliographical references.
Identifiers: IAEAL 26-01818 | ISBN 978–92–0–100726–1 (paperback : alk. paper) | ISBN 978–92–0–100826–8 (pdf)
Subjects: LCSH: Nuclear energy — Management. | Knowledge management. | Manpower policy.
Classification: UDC 621.039:005.94 | STI/PUB/2138

FOREWORD

Nuclear knowledge management and human resources development present both challenges and opportunities for the nuclear industry, reflecting the industry's unique characteristics and an evolving global landscape. Addressing these challenges and recognizing the related opportunities is essential to enhancing the effectiveness, safety performance and sustainability of nuclear operations.

The IAEA plays a crucial role in supporting Member States in these efforts. As a global hub for the exchange, dissemination and preservation of nuclear knowledge and information, the IAEA has organized six previous international conferences on nuclear knowledge management and human resources development, including education and training and capacity building.

As a continuation of this work, the International Conference on Nuclear Knowledge Management and Human Resources Development: Challenges and Opportunities was held in Vienna from 1 to 5 July 2024. The purpose of the conference was to review global developments related to nuclear knowledge management and human resources development; to consider current and future challenges and opportunities; to provide participants with practical solutions that can be applied at the organizational, national and international levels; and to develop and maintain the human resources needed to support safe and sustainable nuclear power programmes.

These Proceedings include the speeches delivered at the high level, plenary and keynote sessions (or their summaries); summaries of all parallel technical sessions and moderated panel discussions; and the full texts of the statements made during the opening and closing sessions of the conference. The accompanying supplementary files, available on-line, include the conference programme and individual papers submitted by the authors. Video recordings of the conference proceedings, presentations and posters presented by the participants are available in the members area of the IAEA Nuclear Knowledge Management Hub.

The IAEA is grateful to all conference participants, in particular the Programme Committee, the session chairs and the technical session rapporteurs, for helping to make this conference a success. Special thanks are extended to E. Pule (South Africa), who served as the conference president, and to M. Uesaka (Japan) and K. Pringle (Saudi Arabia), who served as vice presidents. The IAEA also acknowledges the generous support of all exhibitors, side event organizers and sponsors. The IAEA officers responsible for this publication were P. Diéguez Porras and A. Iunikova of the Division of Planning, Information and Knowledge Management.

CONTENT

1. EXECUTIVE SUMMARY

The International Conference on Nuclear Knowledge Management and Human Resources Development: Challenges and Opportunities, organized by the International Atomic Energy Agency (IAEA), took place from 1 to 5 July 2024 at the IAEA Headquarters in Vienna, Austria. The conference convened over 760 participants from 107 Member States and 9 international organizations, including senior officials, industry leaders, and experts, to deliberate on enhancing the effectiveness and sustainability of the global nuclear sector through strategic knowledge management and human resources development.

The Programme Committee evaluated 450 abstracts submitted by participants and selected 144 for oral presentations and 133 for poster presentations. Ultimately, 123 papers (abstracts) and 72 posters were presented and discussed at the conference. The lists of these presentations are available as Annexes to this publication. Received papers are available on-line in the supplementary files.

The conference opened with address by the IAEA Director General R.M. Grossi and welcoming remarks by the conference President, E. Pule (South Africa), and conference Vice Presidents M. Uesaka (Japan) and K. Pringle (Saudi Arabia).

The conference programme encompassed 2 high-level sessions, 11 keynote, plenary or moderated panel discussion sessions, 24 parallel technical sessions, and 8 side events, structured around four central themes: People, Technology, Alliances and Sustainability.

The following key themes and topics were addressed during the conference:

Holistic Views on Nuclear Energy Policies and Their Impact on Nuclear Knowledge Management (NKM) and Human Resource Development (HRD)

On the first day of the event Member States, including Argentina, China, Egypt, France, Ghana, Japan, Morocco, the Russian Federation, Hungary, Saudi Arabia, South Africa, the United Arab Emirates, and the United States, shared national statements underscoring their commitment to building competent workforces tailored to their unique national contexts. These discussions reaffirmed the IAEA's role in promoting the safe and peaceful use of nuclear technology globally.

PEOPLE: Develop, Empower, Lead

Sessions on the second day focused on innovative training programmes, national workforce development strategies, and global competency models. Discussions emphasized the importance of safety and security culture, effective knowledge transfer, and leadership development. A notable session addressed strategies for retaining talented women in nuclear organizations, highlighting the need for inclusive practices and leadership opportunities.

TECHNOLOGY: Enabling Innovations and Knowledge

On day three participants explored the role of advanced technologies, such as artificial intelligence and digital platforms, in supporting NKM and HRD. Sessions highlighted the potential of AI and large language models in enhancing training and knowledge management, as well as the application of modern tools in various nuclear sectors, including waste management and decommissioning.

ALLIANCES: Engaging Youth Through Global Collaboration.

On the fourth day the conference underscored the significance of international collaboration in engaging the young generation. Discussions centred on university education, educational networks, and strategies to attract young professionals to the nuclear field. Parallel sessions and panel discussions provided platforms for sharing experiences and best practices in fostering global partnerships and youth engagement.

SUSTAINABILITY: Looking Forward, Resilience and Progress

The final day of the conference focused on strategies to maintain and enhance the sustainability of NKM and HRD to support the long-term viability of the nuclear sector. Following the keynote presentations, the plenary session delved into comprehensive strategies for sustaining NKM and HRD initiatives. Discussions highlighted the integration of sustainability principles into organizational policies, the role of leadership in fostering a culture of continuous learning, and the importance of adapting to technological advancements to maintain relevance in the evolving nuclear landscape. By addressing these critical areas, the conference aimed to equip stakeholders with the knowledge and tools necessary to navigate the challenges of the future and to contribute to the safe, secure, and sustainable use of nuclear technology worldwide.

Throughout the conference, e-poster and paper poster sessions provided additional insights into innovative approaches and research findings related to the theme of the day and discussions during the keynote, plenary or parallel technical sessions. Participants had the opportunity to engage with authors, discuss methodologies, and explore potential collaborations.

On the margins of the conference 8 side events were organized by various Member States and organizations, each focusing on specific aspects of NKM and HRD. China's side event highlighted its advancements and initiatives in NKM and HRD, sharing best practices and experiences. France hosted an event on international mobility, discussing strategies to enhance workforce mobility within the nuclear sector. EUROfusion's event delved into NKM and HRD aspects specific to fusion organizations, emphasizing the unique challenges and solutions in this field. The Russian Federation presented its technological innovations in NKM, showcasing recent developments and applications. The ENEN side event focused on the pivotal role of European collaboration in strengthening nuclear competence. The IAEA Nuclear Knowledge Management Section organized a side event discussing its services and assistance to address various Member States' needs in NKM and HRD, while the IAEA Technical Cooperation Department hosted an event on capacity building and technical cooperation. Additionally, the IAEA Department of Safety and Security conducted a side event emphasizing the integration of safety and security considerations into NKM and HRD practices. These side events provided platforms for in-depth discussions, knowledge sharing, and the exchange of best practices among international stakeholders, highlighting the IAEA's commitment to providing Member States with all necessary support and assistance.

During the conference the proSTEM Challenge 2024 recognized five outstanding projects that demonstrated exceptional innovation in promoting STEM education within the nuclear sector. The award ceremonies were organised as a part of the relevant sessions and side event. They celebrated the finalists' contributions, underscoring the importance of fostering a skilled and knowledgeable workforce to ensure the future sustainability of nuclear technology and its applications. The list of proSTEM Challenge 2024 finalists and summaries of the projects are available in the Annex V of these proceedings.

The closing session featured a statement by conference President E. Pule, who reported on the conference outcomes, key themes, and discussions from the week, highlighted main issues and conclusions of the event. The IAEA Deputy Director General M. Chudakov also delivered closing remarks from the Agency's perspective. The session reinforced the collective commitment to advancing sustainable practices in nuclear knowledge management and human resources development, ensuring the future resilience and success of the nuclear industry.

2. OPENING SESSION

Verbatim as delivered[1].

R. M. Grossi
Director General of the International Atomic Energy Agency

Good morning, everybody,

It's very good to open this conference. This very important meeting, for one reason or another had to be postponed a couple of times, but finally, we are here. This is very good because quite clearly, we are in a very interesting moment full of promise, but at the same time with a number of issues, that we as a nuclear energy community will have to address. Nuclear knowledge management and human resource development are natural partners and issues that require Member States and the Agency's joint consideration.

I would like to thank you, Madam Chair, for your presence here, for bringing your unique experience from the private sector and also in public policy development in South Africa. I think, it's very important that we underscore the globality of our effort. I also come from a developing country, and I believe it is very important that we have this approach, which is unique to our industry, with all its differences and diversity. I would like to very briefly offer a couple of thoughts on the most important objectives, as I and we see them. Today nuclear energy is playing and is called to play an even greater role in terms of economic development and also in caring for the global environment.

Times have changed. There is a realization that for us to provide the adequate responses to the needs of the economies of the world, and at the same time the goals of decarbonization, nuclear is needed. But there are issues that we need to address. One very important part of it is the financing and what you are doing as of today is also a key element. We are no longer looking at a decaying industry. We are no longer looking at something that has its days counted, as at some point was the assumption that nuclear would be slowly reducing its contribution to the energy mixes internationally and nationally.

Now we see this has changed. We, in this industry, knew nuclear was needed, and now we see a new global consensus as confirmed at the COP 28 in Dubai, just a few months ago, when nuclear was confirmed as something that needs to be accelerated, not only tolerated, not only being there, but accelerated just as renewable energies. But, of course, this requires knowledge management, and this requires a workforce that is going to be up to the task. It also requires money, investment and political decisions. So, this is extremely important.

What are we supposed to be addressing efficiently? First of all, retention. The workforce in nuclear must stay in nuclear. There will be a need for more people working in this area. What are the particulars? What are the specifics that need to be addressed nationally and internationally to achieve this retention, which is also important and compounded with the fact of the need to replace those who will be retiring? It has become commonplace to say that, because of the lack of investment in nuclear over the past years, we did not have, the regular input of new generations working in nuclear. I hear and I see everywhere I go, every plant I visit, every country I am in, that there is a generation that is about to retire and leave the workforce, and there is doubt as to the capacity of the industry to muster a new generation. So, there

---

[1] Opening remarks have been edited for clarity.

are these two issues of retention and replacement, which bring us to the other very important element of attraction.

An attractive industry creates opportunities. The creation of opportunities is something that must be pursued assertively and with concrete actions. From the point of view of the Agency we are doing this through the schools, through the different activities that we carry out together with academia, together with universities, together with industry, we are trying to provide these opportunities and platforms that are needed for young engineers, young technologists to find interest in nuclear. Importantly, as some of you may know, we are looking at the gender perspective, which is so relevant. And we are doing that not only with nice words, which is the typical approach when it comes to gender issues, but with concrete programmes like the Marie Skłodowska-Curie Fellowship that already has helped more than 500 young women, providing opportunities for them to attain a master's degree. The Lise Meitner programme is designed for women already in the industry, giving them career-expanding opportunities. And we are planning to continue and to expand these programmes, both those that are gender perspective and that that seek to incorporate young talent in general into the nuclear family, so to speak.

So, replacement, retention, attraction, but also, lastly perhaps, development. How is the workforce, the nuclear workforce, to be efficiently developed, when we are incorporating also new technologies, new designs, new ways to do nuclear, including with new alternatives and opportunities, for example, in modularity for reactors, as you know, or even fusion which the Agency is looking at so actively. So, Madam Chair, dear friends, this conference looking at the human factors of our industry is absolutely essential. Here, we are going to be looking at what luckily continues to be the essential element, which is us, the people. The technology and the money and all these things are, of course, essential, but we need to know and to understand better how to advance nuclear, and it's the women and the men in nuclear that are going to be doing this for future generations.

So, let me commend you for this very important effort. With the secretariat team and scientific secretariat of the conference we are going to be providing you, Madam Chair, and all the participants with the support you need. Be welcome at the IAEA. Be welcome at this very important conference.

Thank you very much.

As prepared for delivery.

E. Pule
President of the Conference, HR Executive of Eskom Holdings SOC Ltd., South Africa

Good morning, ladies and gentlemen,

Dear colleagues,

My name is Elsie Pule. I am the HR Executive at Eskom Holdings SOC Ltd, in South Africa.

It is my honour to serve as the President of the 2024 IAEA's International Conference on Nuclear Knowledge Management and Human Resources Development, the first in the last two decades to combine these two pivotal topics.

I am delighted to share this podium today with Mr. Rafael Mariano Grossi, the Director General of the International Atomic Energy Agency. I would like to express my gratitude for his inspiring speech to all of us. Thank you also to the IAEA official who have made the needed work to make this conference possible.

It is a pleasure to welcome you to the international conference on NKM and HRD.

Effective nuclear knowledge management is crucial for success in all industries, and especially in the nuclear sector. A nuclear power programme requires a long-term commitment of people and resources. It typically spans several generations of professionals, underscoring the importance of sharing and maintaining specialist knowledge both within the minds of the involved professionals and within the organizations.

According to the IAEA PRIS database, currently 416 nuclear power reactors are in operation in 31 countries, 25 are in suspended operation, and 59 more are under construction.

With its ability to provide stable base-load electricity and emit very low levels of greenhouse gases, nuclear power will be a crucial component of the solution to help countries grow their economies while mitigating the impact of climate change.

Additionally, the use of nuclear technologies in medicine, agriculture, food security, and more underscores the critical role of nuclear power in ensuring sustainability in the coming decades.

Acquiring, retaining, managing, and transferring technical knowledge are major challenges both in countries with established programmes and in those termed as newcomer countries by the IAEA. Attracting a new generation of future professionals to engage in current and new nuclear technology developments, while an entire generation of experts is beginning to retire, poses a dual challenge.

Even in countries phasing out nuclear power, it is recognized that preserving critical knowledge is essential to ensure responsible decommissioning and environmental remediation of sites.

Engaging the new generations of scientific and engineering talent, with special attention to women, utilizing emerging technologies, and fostering international collaboration and alliances are paramount to ensuring the future availability of highly experienced staff. They will assume the responsibility for the safe, secure, and sustainable operation of nuclear facilities and the development of new nuclear technologies in the coming decades.

Ladies and gentlemen, to address these challenges, the IAEA has curated a compelling programme for this conference, encompassing topics such as nuclear knowledge management, human resource development, nuclear education, new information management technologies, and international alliances over the course of five days.

We anticipate expert participants from all corners of the globe, providing invaluable opportunities for nuclear managers, knowledge management specialists, and decision-makers to exchange experiences and lessons learned.

The nuclear sector continually requires the cultivation of knowledge and talent. Managing and developing expertise in nuclear technologies is a broad, demanding, and complex endeavour. Sharing your insights with the international community in events like this one can yield new opportunities to enhance our current practices and tackle emerging challenges.

I wish you productive discussions in the days ahead and look forward to learning and sharing the outcomes of this conference.

Thank you very much.

Verbatim as delivered.

M. Uesaka

Vice President of the Conference,

Chairperson of the Japan Atomic Energy Commission (JAEC), Japan

I am the chairperson of Japan Atomic Energy Commission.

It's my great honour to serve as a Vice President of this important IAEA conference.

I used to be a professor at Nuclear Professional School, University of Tokyo till 4 years ago. I had also been a member of IAEA INMA (International Nuclear Management School) and a chairperson of the Japan-IAEA Nuclear Energy Management School.

High quality NKM and HRD are crucial in the nuclear sectors in the world and Japan.

I would like to continue to contribute to these important tasks.

Thank you very much.

Verbatim as delivered.

K. Pringle
Vice President of the Conference, Director of Human Capacity Building
of the King Abdullah City for Atomic and Renewable Energy, Saudi Arabia

Good morning, everyone.

It's a pleasure to be here. I've heard so many interesting remarks this morning from the Director General and our Chair and our Vice President. What I am most excited about is the opportunities for us to get to know each other better, to learn from each other. Whenever I come away from conferences like this, I really value the friendships that I've created, the new connections and networks that I've created, and what I see here is a room full of people that have a lot to share and a lot that we can all benefit from.

I'm looking forward to speaking later about Saudi Arabia's national position and also our national strategy. Many of the points, that I will cover, were actually covered by you, Director General Grossi, talking about retention, attraction, talking about infrastructure, how do we make sure, that we build a sustainable capacity within the country. I've worked with King Abdullah City for Atomic and Renewable Energy for 12 years. I'm the Director of Human Capacity Building.

We worked on first strategy for nuclear and renewable, for human capacity building, and just recently, last year, His Royal Highness, Minister of Energy, asked K.A. CARE to take the responsibility of working cohesively across the whole entire energy sector. So, we're now working on building a workforce planning unit and then market intelligence unit to look at how we're going to build capacity, not just for nuclear, but how we can benefit from oil and gas, petrochemical and hydrogen and carbon capture, and build a comprehensive strategy for human capacity building for the Kingdom. So, we're moving into an interesting territory and it is really an interesting time for us to talk with each other and learn from each other on how we can build these human capacity building strategies together.

It's an honour to be here today as the Vice President. I hope I'll be able to add value, and I hope as well that this would be enjoyable and beneficial conference for all.

Thank you so much.

3. HOLISTIC VIEW ON NUCLEAR ENERGY POLICIES AND THEIR IMPACT ON NKM AND HRD: GLOBAL CHALLENGES AND NATIONAL ALTERNATIVES

As prepared for delivery.

J. Gadano

Vice President, Nucleoeléctrica Argentina S.A., Argentina

It is a privilege to address you today on the topic of nuclear knowledge management policies in Argentina. Our country has a long-standing commitment to nuclear science and technology. Over the past decades, we have implemented a comprehensive framework to ensure the effective management, preservation, and transfer of nuclear knowledge.

Why? Because in 2006 we suffered the consequences of not having worked harder to build a robust knowledge management structure. An example: when we decided to restart the Atucha II Nuclear Power Plant project (after 20 years of paralysis) we found that much of the knowledge had disappeared. This allowed us to build awareness of the importance of keeping knowledge management updated.

This framework is supported and guided by the principles set forth by the International Atomic Energy Agency (IAEA).

This conference is a special occasion this year, as we not only celebrate the 20th Anniversary of the Joint ICTP (International Centre for Theoretical Physics)-IAEA Nuclear Knowledge Management School, a vital tool for member states. Just as a reminder: the school basically consist in a certificate course providing specialized education and training on development and implementation of knowledge management programmes in nuclear science and technology organizations. It is intended for young professionals in current or future leading roles in managing nuclear knowledge.

This fact also serves to point out nearly 20 years of Argentina's commitment integrating knowledge management into nuclear science and technology. In 2003, we Argentina received, through our Atomic Energy Commission – CNEA, the first expert mission by the IAEA to establish a formal knowledge programme with the support of the agency.

First and foremost, we have to emphasize that the core of Argentina's nuclear knowledge management efforts is a robust institutional framework. The CNEA (or National Atomic Energy Commission), with over 70 years of leadership in nuclear R&D activities, plays a crucial role as a Technical Support Organization (TSO) in capturing and preserving nuclear knowledge through education, training, and research. Additionally, the Nuclear Regulatory Authority (ARN), our independent regulatory body, founded 70 years ago and working as an independent institution for 30 years, complements these efforts by ensuring our knowledge management practices meet the highest safety standards. Nucleoeléctrica Argentina S.A., as the utility of NPPs, focuses on retaining critical operational knowledge, enhancing safety and security, and supporting continuous improvement and innovation.

Let me now to mention the 1st Argentine School of Nuclear Knowledge Management, carried out in 2017 at the Central Headquarters of the National Atomic Energy Commission (CNEA), in which INVAP, Nucleoeléctrica Argentina, FUESMEN, the Nuclear Regulatory Authority (ARN) and the Undersecretariat of Nuclear Energy, under my charge at that time, also participated.

With the support of the IAEA and the Latin American Network for Education and Training in Nuclear Technology (LANENT), more than 50 specialists participated in this activity, which was the kick-off of current Knowledge Management programmes in the different institutions of the Argentine nuclear sector.

Allow me to quote my own words as Deputy Secretary of Nuclear Energy at that time: "The nuclear sector, and specifically the CNEA, has a very valuable group of people, who contributed significantly to Argentine development. But at this moment we have the responsibility of transferring knowledge to new generations and young people, so that knowledge remains in the institutions, beyond the people who temporarily make up them." I remember that we also emphasized that "knowledge is something that is built collectively."

Our National Policy on Nuclear Knowledge Management focuses on developing and implementing tailored programmes for nuclear science and technology organizations and public companies. This includes IT methodologies and practices, organizational culture and human resource development for knowledge preservation and sharing. These factors represent some of the primary challenges we face today in the nuclear sector, both nationally and globally.

The IAEA underscores the importance of a robust institutional framework for effective nuclear knowledge management. By adhering to its guidelines, Argentina ensures that its institutions are well-equipped to manage and preserve critical information and know-how. This is especially important as our National Nuclear Programme evolves (including life extension of NPPs) designing and constructing our own Small Modular Reactor (SMR) and research reactors and investing in other productive nuclear activities and facilities.

Our commitment to education and capacity building has been evident since the 1950s, starting with the establishment of the Balseiro Institute, a regional benchmark for physics and nuclear engineering. Building on this success, we have created additional Nuclear Academic Institutes and developed academic programmes in collaboration with public universities, focusing on nuclear safety and various nuclear technology applications, including energy, medicine, agriculture, and the environment.

Through strategic partnerships with these academic institutions and their specialized training programmes, we are nurturing the next generation of nuclear experts and technicians. This effort is a cornerstone of our nuclear policies and their sustainability over time. However, we recognize that the challenge extends beyond merely preserving and transferring knowledge.

Engaging new generations is crucial for maintaining continuity and enhancing our capabilities. A strong commitment from both the public and private sectors in developing new technologies, state promotion of new investment projects, and an academic sector aligned with professional demands are crucial issues for decision-makers to address.

Our pledge to excellence in nuclear knowledge management is reinforced through active participation in global regimes, multilateral forums, and international conferences, workshops, schools, and training programmes. These engagements enable our professionals to exchange expertise and learn from experiences, continually enhancing our knowledge management practices. We are also focused on incorporating new strategic tools, particularly Information Technology (IT).

In today's digital era, IT is indispensable for managing knowledge in general, and nuclear in particular. We leverage digital repositories, databases, and collaborative platforms to efficiently store, share, and access nuclear information, crucial for enhancing knowledge dissemination and ensuring the durability of critical information. We believe that ongoing efforts in this area are essential to further enrich our organizational culture and community in this domain.

It is also important to highlight some challenges that lie ahead. Faced with the growth of the global nuclear industry (a phenomenon that we estimate will increase in the coming years) and the projects underway in our country (the completion of the CAREM and RA-10 reactors, and the LTO of Atucha I power reactor), the importance of improving knowledge management in our organizations is evident. An entire generation is retiring and, although we have managed to improve the average age of our staff, knowledge management has not always been ideal, and we will have to face challenges in this field.

Likewise, new technological developments, such as AI, force us to build knowledge and training networks not only within the nuclear sector but also with the university system.

In conclusion and being clear about the challenges we face, Argentina's comprehensive approach to nuclear knowledge management underscores our persistent commitment to maintaining high levels of expertise and safety in our nuclear programmes. Through strategic education, rigorous preservation and transfer practices, international collaboration, ongoing research and development, and the effective use of technology guided by IAEA principles, we ensure that our nuclear knowledge is effectively managed and utilized for the benefit of our industry and society.

Verbatim as delivered.

H. Li
Deputy Director General for HRD, China Atomic Energy Authority (CAEA), China

Good morning.

It's my honour to attend this conference and share with you China's good practices and experiences in promoting the sustainable development of nuclear energy and the human resources.

Now I will speak in Chinese, and the interpreter will translate into English.

As the world's largest developing country, China takes nuclear energy as an important option to ensure the energy security, build a new energy system, and achieve the goal of carbon peak and carbon neutrality. China has been following a rational coordinated and balanced nuclear security strategy put forward by the Chinese president Xi Jinping and formulated and implemented policies and measures for the active, safe, and all state development of nuclear power.

Currently, there are 57 nuclear power units in operation in Mainland China. And the scale of nuclear power units under construction has been leading in global for years. The units in operation and under construction maintain good safety performance with no instance of INIS level 2 or above. In addition, China has been actively promoting the application of the nuclear science and technology in the food and agriculture, human health and environmental protection.

The sustainable development of peaceful use of nuclear energy could only be achieved with the support of high level professional and successive talents. China attaches great importance to the human resource development in the nuclear sector and regards talents as the primary source and the fundamental for promoting the innovation of nuclear science and technology, as well as the sustainable development of nuclear energy.

At first, we built a virtuous circle among the education, technology, talent, and the working mechanism for the talent development, induction, service, and turnover. The system of nuclear technological innovation and talent cultivation is established where government, universities, research institutes, and industries could work together to promote each other. A large number of the landmark achievements have been, made such as the self-developed 3rd generation nuclear technology, HPF 1 000, the world's first 4th generation high temperature gas cooled reactor demonstration project, experimental and demonstration fast reactor, and fusion experimental device, China loop 3. China has a large number of high-level scientists, engineers and skilled professionals in the nuclear sector.

Second, we optimized the discipline setting and talent training mode in universities, laying a solid foundation for the nuclear energy development. At present, China has set up the nuclear related disciplines and majors with distinctive, comprehensive, and diverse ranges in more than 70 universities across the country. The talent's development mechanisms have been established through the provincial versus ministerial cooperation and university versus industry cooperation. Every year, there are 10 000 graduates delivered to the nuclear industry. It forms a steady stream of the talent's backup for the nuclear energy development.

Third, we combine the industry with education and theory with practice, developing the high-level skills talents for the sustainable development of nuclear energy. With the resources of nuclear industry, a number of national level training centres for the high skills talents and national level skill studios have been set up. Through the basic teaching, the practical training, and skill training, a large number of skilled professionals have been cultivated throughout the entire nuclear industry chain, including the construction, the equipment manufacturing, the nuclear facility operation, and maintenance.

Fourth, we carry out the nuclear science dissemination and public communication activities through multiple approaches to create a positive social environment for developing talents. Since the year of 2013, 12 sessions of the Charming Light Nuclear Science Dissemination and Public Communication Activities have been held nationwide. Through the knowledge competition, the summer camp, the academic experts entering the campus, the nuclear science education are carried out in the universities and middle schools, and the nuclear knowledge is popularized to the society and the communities. A good atmosphere of knowing, understanding, and supporting the nuclear energy has been gradually formed in the society.

Ladies and gentlemen,

This year marks the 40th anniversary of China's accession to the IAEA. Over the past 4 decades, China has been the only country with the ongoing continuous nuclear power construction. We have built various types of global mainstream nuclear reactors such as the EPR, VVER, and AP1000 in China. We have formed our all-nuclear power brand and a complete nuclear industry chain. We accumulate rich technology, knowledge, and talent backup in the design, construction, operation, and maintenance of nuclear power.

Meanwhile, China actively shares its experience in the nuclear energy development with other countries and provides support and assistance to developing countries in the human resource development and capacity building of nuclear energy. We have set up programmes such as the Atomic Energy Scholarship of the Chinese government, and the international master's programme in nuclear engineering and management. So far, the programmes above have educated over 400 master and doctoral candidates in the nuclear field for more than 30 developing countries. In cooperation with the Agency, we have set up international training centres for nuclear power capacity building and a number of collaboration centres in nuclear energy and nuclear technology application. The China-IAEA Nuclear Energy Management School has been jointly implemented as well. Through IAEA's technical cooperation programmes thousands of experts from developing countries come to China for scientific visits, technical exchanges and short-term trainings.

Facing the future, China is ready to work alongside with the Agency and the like-minded partner countries to jointly implement the global development initiative and the global security initiative, actively aligned with the Agency's flagship initiatives such as Atoms4NetZero, Rays of Hope, and Atoms4Food. China is willing to provide more public goods and services in the human resource divestment, the knowledge dissemination, and capacity building for the peaceful use and safe development of nuclear energy for all countries so as to promote the sustainable development. We will work together with all Member States and the secretariat to achieve the goal of "Atoms for Peace and Development", and build an open, inclusive, clean, and beautiful world that enjoys the lasting peace, universal security, and common prosperity.

Thanks.

As prepared for delivery.

S. Sharshar
General Manger of Public Relations and Public Acceptance
Nuclear Power Plants Authority, Egypt

El Dabaa NPP Project

Nuclear Knowledge Management and Human Resource Development Aspects

1. Human Resources Development Strategy for El Dabaa Nuclear Power Plant

The Nuclear Power Plants Authority (NPPA) of Egypt is responsible for the safety and operational management of the nuclear facilities it establishes. This includes overseeing activities related to design, licensing, quality assurance, security, supply, manufacturing, construction, and the management of employees and contractors, as well as future operations.

2. Strategic Objectives

Egypt developed a comprehensive "Egyptian Human Resources Strategy to Launch and Sustain the Nuclear Power Program" early in the project's conception. The strategic objectives include:

- Long-term Governmental Commitment: Ensuring sustained governmental support for the nuclear power programme and human resources development;
- Roles of Programme Partners: Clarifying and optimizing the involvement of all stakeholders;
- International Cooperation: Partnering with international organizations, consultants, and nuclear plant suppliers for training purposes;
- Defining Required Competencies: Clearly defining the necessary activities and competencies for the programme;
- Organizational Structure: Modifying structures to meet the needs of the nuclear power programme.

3. Challenges and Vision

Constructing the first nuclear power plant is a significant challenge and the need to develop the necessary skilled personnel for its operation and maintenance is a key element of this challenge. The nascent nature of Egypt's commercial nuclear power programme necessitates the recruitment and training of a large contingent of graduates and experienced professionals from other technical industries.

The NPPA aims to staff the organization entirely with Egyptian nationals, in line with its vision of contributing to sustainable development and leveraging nuclear power for Egypt's future. Achieving this vision depends on developing skilled human resources who understand the complexity of nuclear technology, adhere to high safety standards, respond promptly to incidents, and minimize human error.

4. Education and Training

Egypt's higher education sector, including 18 universities and numerous technical colleges, offers degrees in nuclear engineering and other relevant technical fields. The Ministry of Education has established the Advanced Technical School for Nuclear Energy Technology in El-Dabaa, and the Egyptian Russian University has a Nuclear Power Station Engineering Department. Egypt's two research reactors, procured from Russia and Argentina, support the nuclear industry.

The NPPA has sent personnel to IAEA workshops and training courses, as well as nuclear energy programmes in Korea and Russia. Collaborations with the United States have also facilitated nuclear energy workshops in Egypt.

The EPC contract for the construction and commissioning of El Dabaa NPP includes comprehensive training programmes for close to 2000 NPPA personnel and also provides for the construction of two full-scope simulators, in order to prepare for commissioning and operation. Additionally, the NPPA has contracted an organization to support El Dabaa NPP staff in operation and maintenance during the initial years, providing oversight, advice, and on-the-job training.

5. Integrated Management System (InMS) and Human Resources Development

The NPPA has developed its Human Resources Development (HRD) function in line with IAEA safety standards, the World Association of Nuclear Operator's (WANO) 'Roadmap to Operational Readiness', and national regulations. The InMS incorporates a systematic approach to training (SAT) for developing and maintaining personnel qualifications.

Cognisant of the nascent nature of Egypt's commercial nuclear power programme, NPPA recognises that cross-functional knowledge is essential. Many employees receive training in multiple areas, fostering an integrated knowledge base.

6. Knowledge Management

Knowledge Management (KM) is crucial for NPPA's success, aiming to accumulate relevant knowledge and continuously improve performance. The KM subprocess within the InMS includes:

- Identification and Prioritization: Identifying key knowledge domains important for NPPA's goals;
- Knowledge Capturing: Bringing data and knowledge into the organizational knowledge base;
- Knowledge Development and Retrieval: Maintaining, storing, and retrieving knowledge to preserve valuable information;
- Knowledge Sharing: Encouraging an organizational culture that rewards knowledge sharing.

7. Consultancy Support and Operational Readiness

The NPPA employs consultancy firms with expertise in nuclear power plant construction and operation to assist in skill transfer and mentoring. Training includes courses on HR fundamentals in the nuclear industry, benefiting from the consultants' experience in setting up HR systems for nuclear programmes.

8. Operational Readiness

The NPPA is developing plans to implement a programme of operational readiness in line with WANO's 'Roadmap to Operational Readiness'. These plans are being developed by NPPA staff with consultant support, with the aim of ensuring that the necessary processes, skills, training, and authorizations have been, or will be, attended to. The milestone of commissioning testing for Unit 1will be a key measure of the effectiveness of the NPPA's HRD function.

9. Conclusion

The Egyptian HR Strategy for the El Dabaa NPP focuses on developing a skilled workforce through strategic objectives, international training, integrated management systems, and effective knowledge management. By leveraging international cooperation and consultancy support, the NPPA aims to achieve operational readiness and sustain the nuclear power programme for Egypt's future.

Thank you.

As prepared for delivery.

G. Bouyt
Director for Nuclear Energy at the French Ministry for Economy, France

Human Resources for Nuclear Energy in France

Madam Chair, Dear Colleagues,

It is an honour for me to be here with you for this conference on nuclear knowledge management and human resources development, as nuclear energy is getting a new momentum worldwide.

This is particularly the case in France. I will try to explain why, while replacing this question into the broader context of French energy policy.

The French nuclear industry is facing an ambitious challenge for the next 30 years, since it has to deliver the biggest industrial projects for the country:

- The construction of 6 new EPR2 reactors;
- The long-term operation of the existing nuclear fleet, while securing the highest possible production availability;
- What is more, the potential renewal of facilities for fuel reprocessing and MOx fuel production.

The French nuclear industry consists in:

- More than 3,000 companies;
- Annual sales of around 50 billion euros;
- Around 220,000 people, including direct and indirect jobs – that is around 7% of all industrial jobs in France. Around two-thirds of these jobs are highly skilled.

We may observe that there is a limited number of big nuclear companies in France, among them utility EDF, its subsidiary Framatome for fuel fabrication and nuclear boiler conception and fabrication, and fuel management company Orano. Then there are thousands of small and medium sized companies, which are not exclusively dealing with nuclear – they could be involved for instance in the aviation of health sectors.

For big and small companies, maintaining competence across the whole nuclear industry needs sufficient long-term visibility, for companies and people to invest in building competence on the longer term. And the nuclear industry may be the one with projects having the longest duration over time.

Key nuclear competences enter a new development phase.

In France, our new nuclear programme follows the Flamanville 3 construction, before which we had around 20 years without nuclear new built. All the lessons must be learnt from previous projects, technical ones, but also as regards competences: it is necessary to secure the highest standards in welding, piping, but also in overall complex project management.

There is a qualitative challenge in skills, there is also a quantitative challenge in skills for the completion of a vast volume of projects in the next decades. But we know that quantity and clear time perspectives on the long term will help build and maintain quality.

Both the French Government and industry are involved in building and implementing a comprehensive plan for adequate competence in the nuclear business.

The public investment plan France 2030 dedicates more than 1 billion euros to support the nuclear sector, and in addition, 2.5 billion euros are identified for competence building initiatives across all industrial sectors.

GIFEN, the French Nuclear Industry association, launched in April 2023 the MATCH project which first aims at assessing the adequacy between, first, the nuclear workload and, second, skill resources over the next 10 years:

- From this analysis follows the recruitment timetable for the period 2023-2032, category by category of specific skills;
- The MATCH report also stresses the importance of continuing the implementation of performance plans in every nuclear company with the aim of achieving and maintaining operational excellence.

The MATCH project answers a request from the Government and will be updated on an annual basis. It will also be complemented with a special focus on needs for NPPs with French content abroad.

The industry also gathered to launch the University of Nuclear Skills, which is structuring the academic and professional training resources in France for the nuclear business. The University of Nuclear Skills produced a plan with the objective to enable 10 000 good recruitments in the nuclear sector each year for the next decade.

Madam Chair, Dear Colleagues,

To conclude, I would like to share some thoughts on success factors for the nuclear sector, the performance of which essentially relies on human resources:

- Of course, the first key lever is the attractivity of the nuclear sector. Perceptions are evolving in the general public, slowly but strongly, to identify nuclear as a key asset for lowering carbon emissions and fighting climate change. This is the challenge of our century, and this speaks to all of us, especially younger ones. What is more, industrial jobs pay well and are stable when the industry has long term perspectives. This is what we should hear more in middle schools and high schools.
- A huge potential is also to attract more women in nuclear. Mentorship could be a strong tool for this. The example of amazing careers led by women in previous decades, when such paths were even harder to achieve, could contribute to reverse stereotypes and enable the nuclear sector to benefit more from fantastic women who could join. That is what we try to encourage in companies, and at institutional level, with conferences, visits and mentoring.
- A key point is also to build the bridge between experienced nuclear professionals that have a lot to teach and share and the younger generations which want to embark. I believe in the training capacity of big nuclear companies, but also in the special role which start-ups can play – and they are now a lot flourishing in France –, to attract younger, brilliant, digital ones in close contact with seasoned experts on each type of technology, even fast neutrons or fusion.
- I am also convinced that competence building should be thought directly at regional and international level, with countries which pursue similar nuclear projects. We need to find ways to share competence and people between projects when they phase well, and progressively build a strong nuclear competence basis at regional level.

France is fully committed to reach its Paris Agreement objectives and we need a strong international nuclear workforce to succeed!

I thank you for your attention.

Verbatim as delivered.

S. Dampare
Director General of the Ghana Atomic Energy Commission (GAEC), Ghana

Session chairpersons, distinguished colleagues, ladies and gentlemen,

It is an honour for me to address you today. I would like to thank the organizers of the conference for the invitation, and I bring you warm greetings from Accra, Ghana.

There's a popular saying that knowledge is power. However, knowledge must be effectively managed to get its power fully. We need technology processes, procedures, and most importantly, people to manage knowledge.

Thus, knowledge management demands competent human resources with the proper skills, attitude, perspectives, and most importantly, familiarity with the subject matter. With this in mind, Ghana appreciates the IAEA's tremendous role over the years in showing leadership when it comes to initiatives aimed at enhancing nuclear knowledge management and human resources development for peaceful nuclear technology applications. The Agency's activities through bilateral regional and regional support initiatives, technical infrastructure installation and upgrades, as well as advisory services in NKM and HRD are felt globally. Ghana finds this conference most approaching as it affords the international community a good platform to share and discuss issues of common interest pertaining to NKM and HRD. This conference could not have come at better time as Ghana advances its nuclear power programme. I'm happy that some Ghanaians are here at this very important conference to engage and learn.

Ghana decision to develop a nuclear power programme is influenced partly by the country's long-term vision of creating a resilient energy infrastructure that is climate adaptable and environmentally sustainable. The country also has an ambitious national development plan that necessitates a secure, affordable, and diverse national energy portfolio to mitigate the impact of rising electricity demand, limited hydro resources, and the potential depletion of gas reserves on the national development agenda. It is worth noting that since 2012, Ghana's intention to develop nuclear power for national development has been clear. At the moment, nuclear infrastructure and its associated issues are being developed and addressed according to the IAEA milestone approach, and the programme is currently in phase 2 of the 3 phase milestones approach.

It must be noted that the initial effort to develop national nuclear power infrastructure in the early 1960s suffered setbacks along the line for a variety of reasons, including lower electricity demand at the time. Research into nuclear technology and its applications continued. Thus, the Ghana Atomic Energy Commission, established in 1963 by an act of parliament as the national institution in charge of the promotion of the peaceful applications of nuclear science and technology, has continued its research into the technology.

Over the years, Ghana has strategically expanded its activities to align with global nuclear developments while meeting national health, agriculture, economic, educational, environmental, and social needs.

Ladies and gentlemen,

A great boost to Ghana's overall nuclear technology programme, which encompasses various sectors both power and non-power applications, is the establishment of a graduate School of Nuclear and Allied Sciences (SNAS) in 2006. As a good example worthy of emulation, SNAS is the product of the collaboration between the Ghana Atomic Energy Commission and the University of Ghana with the support of the IAEA. Ghana through this arrangement, which is anchored around as the primary national institution for nuclear technology issues, has worked to enhance nuclear knowledge management and human resources development in the country. It has also partially met the human resources development needs of some African countries, following SNAS' attainment of regional designated centre (RDC) status, offering targeted training programmes in nuclear technology and applications. For about 17 years now, SNAS has successfully graduated 47 Doctors of Philosophy students. That is 41 males and 6 females, and 688 Master of Philosophy students in several nuclear related academic programmes under the departments of nuclear engineering, nuclear sciences and applications, nuclear agriculture, medical physics, and nuclear safety and security. Additionally, over the years, SNAS has hosted an IAEA Postgraduate Educational Course in radiation protection and safety of radiation sources (PGEC). The PGEC staff spans 5 to 6 months for each session and has trained 11 cohorts of fellows from all-over English-speaking Africa and has produced 227 Ghanaian and Africans with enhanced knowledge and competencies in radiation protection and nuclear safety.

Distinguished chairpersons and participants,

Dwelling on the theme of this conference, I would like to focus on national efforts to improve nuclear knowledge management and human resources development thus far. In Ghana, GAEC is taking the lead in implementing a nationwide NKM and HRD strategy to meet the needs of the National Nuclear Power Programme. The first step toward this goal was the conduct of a gap analysis across all aspects of the national nuclear infrastructure. Focusing on NKM and HRD, some of the gaps identified included:

(1) Limited public education on nuclear issues;
(2) Low level of public interest in nuclear applications;
(3) Limited level of awareness of the benefits of nuclear technology;
(4) Limited level of nuclear science concepts in science, technology, engineering, and mathematics, STEM subjects at the various levels of education;
(5) Limited research in nuclear science and application at the higher level of education and research institutions.

Ladies and gentlemen,

With the foregoing in mind, permit me to highlight some of the solutions the country has fashioned towards addressing this NKM and HRD gap.

1. Enhancing public perceptions. The nationwide public perception survey on nuclear technology found that people have preconceived notions about the technology and its applications. To address this, the Ghana Nuclear Power Programme Organization (GNPPO), established a stakeholder industrialization, that has developed public education programmes. For example, one of these programmes includes meeting with editors and senior management from various media outlets in the country to discuss the appropriate ways to present news about nuclear power to the public to produce the desired outcome.

2. Integrating nuclear science into STEM curricula. It was recommended that nuclear science and technology should be integrated into STEM curricula at all levels of education. Even though this requires a significant shift in the national educational policy, policy makers have embraced it.

3. Facilitating the Ministry of Education's commitment to Technical and Vocational Education and Training (TVET). The Ministry of Education in Ghana is committed to Technical and Vocational Education and Training (TVET) as a vital enabler in the national industrialization drive. Consequently, GAEC has established a centre for TVET that focuses on offering practical training experience to develop skilled human resources. In addition, training of artisanal labour force and craftsmen is key to our nuclear programme success. Thus, the establishment of the Regional Clean Energy Training Centre under the auspices of the US Department of Energy, and the recently signed a MoU towards the

establishment of SMR full scope simulator and welding training facility under the US Development of States' Foundational Infrastructure for Responsible Use of Small Modular Reactor Technology (FIRST programme), – are all meant to develop human resource in Ghana.

4. Involvement of the country's policy makers in the national nuclear programme. Politicians, opinion leaders, civil society organizations and other stakeholders are involved in nuclear programme activities through meetings, scientific visit, and public discussions. The various organizations of the nuclear power programme, namely Nuclear Regulatory Authority (NRA), Nuclear Power Ghana (NPG) and Nuclear Power Institute of GAEC are making efforts to develop human resources by identifying crucial skills needed for various roles in the nuclear power programme and conducting targeted training programmes based on the identified skills.

Thankfully, our international partners have been helpful by providing expertise, tools, and resources. As an embarking country, we create and acquire knowledge in the development of our national nuclear infrastructure as we go along. We find the IAEA resources to be essential in developing and implementing strategies to adequately capture and share the nuclear knowledge.

Chairpersons,

The world is fast moving to mainstream artificial intelligent (AI) and augmented reality AI in many facets of life. Thus, we understand that we also need to develop AI tools to improve the potentially revolutionize nuclear knowledge management and human resources development. Thankfully, some tools are already being used to facilitate knowledge sharing via intelligence search platforms, streamline knowledge transfer, and provide faster access to critical knowledge.

Distinguished chairpersons,

Concluding, let me reiterate that the importance of long-term nuclear knowledge management and human resources development cannot be overemphasized. For us in Ghana, we have seen the benefits of deliberate planning and concerted efforts and how they can influence and affect the pace of any nuclear programme, especially for newcomer countries. We have witnessed particularly how many of the key personnel and decision makers at the various stakeholders organizations driving the nuclear power agenda in Ghana, such as Ghana Nuclear Power Programme Organization, Nuclear Power Ghana, Ghana Atomic Energy Commission, and the Nuclear Regulatory Authority have benefited from nuclear knowledge management and human resources development activities and initiatives, such as SNAS academic programmes and the IAEA fellowship, workshops and technical meetings, advisory services, etc. Furthermore, some of our personnel have had the opportunity to train and intern at facilities in advanced and mature nuclear countries, and we believe this coaching and mentoring will go a long way to enhance our nuclear programme for both power and non-power applications.

Ladies and gentlemen,

Despite the successes shown for us in Ghana, we recognize the challenges of nuclear knowledge management and human resources development. For example, the limited number of females in the nuclear industry and its indirect effect of limited female role models for young ladies is a point for reflection. Nonetheless, we also recognize opportunities NKM and HRD presents to the nuclear industry. I'm hopeful that the nuclear industry will continually strengthen its nuclear knowledge management systems to effectively capture and share critical knowledge. We need to implement the necessary training and educational mechanisms to improve the development of a competent nuclear workforce for the safe and sustainable nuclear industry for the benefit of all.

Thank you for your kind attention.

Verbatim as delivered.

M. Uesaka
Chairperson of Japan Atomic Energy Commission (JAEC), Japan

New Builds and Long Term Operation in Japan

Regarding Japan's current nuclear policy and the status of its nuclear power sector, in 2021 the Ministry of Economy, Trade and Industry issued the Basic Energy Plan up to 2030 and aims to produce the revised policy up to 2040 by March 2025. Nuclear is expected to account for 20–22% of the power generation mix by 2030. The Basic Policy for Nuclear Energy of the Japan Atomic Energy Commission (JAEC) is a compass for nuclear policy. It is revised every five years and forms the basis for new nuclear policy and the revision of nuclear-related laws. The new Basic Policy was approved in February 2023.

Since the Fukushima Daiichi accident, 12 nuclear power plants (NPPs) have been restarted, all of which are pressurized water reactors located in the west of the country. We expect a wave of restarts to move east and north this year, aiming to restart four boiling water reactors: Unit 2 of Onagawa NPP, Unit 2 of Shimane NPP and, hopefully, Units 6 and 7 of Kashiwazaki-Kariwa NPP, which is operated by the Tokyo Electric Power Company (TEPCO).

In 2023, the Government decided to extend the NPP operating term from 40 to 60 years, which will allow existing capacity to be maintained up to 2040, after which point new builds will be needed.

In relation to Fukushima Daiichi NPP, in August 2023 TEPCO successfully performed the first discharge of water treated using the Advanced Liquid Processing System, from a discharging point 1 km from the coast. The sixth discharge is currently being carried out very safely. We highly appreciate the great support and help from the IAEA, and especially from the Director General, who has visited the NPP several times. TEPCO performs daily monitoring of the levels of tritium and related radionuclides in the treated water to ensure its safety for discharge and publishes the results on its website a month later in both Japanese and English. In addition, TEPCO will conduct a trial retrieval of nuclear fuel debris from Unit 2 of the NPP later in 2024. Once the retrieved material has been carefully analysed, TEPCO will move towards regular retrieval from Unit 2, followed by Units 1 and 3 in the near future.

Japan continues to develop next generation NPPs, including advanced light water reactors, small modular reactors (SMRs), fast reactors, high temperature gas cooled reactors and fusion energy. We have a road map up to 2050 for developing these reactors as part of our Green Transformation and will invest 1 trillion yen in that work over the next 10 years. Multipurpose use is a vital aspect of innovative reactor development, offering not only power generation in conjunction with fluctuating renewable energy sources, but also heat, hydrogen and radioisotope production. TANDEM, the European Union's project to develop SMRs for a safe and decarbonized European energy mix, is one such effort in this area.

The JAEA has recently started work to reduce the hazardousness of spent fuel by incorporating a fast reactor fuel cycle into the current fuel cycle. The proposed system will recycle plutonium and long lived minor actinides, to reduce radiotoxicity of the disposed waste to that of natural uranium in about 300 years, compared to about 100 000 years with direct disposal of spent fuel.

In 2022, the JAEC issued an action plan to promote the production and utilization of medical radioisotopes with a view to soon producing actinium-225, astatine-211and molybdenum-99 /technetium-99m domestically to treat prostate cancer and Alzheimer's disease, and is now following up on the plan with the relevant universities, institutes, hospitals and organizations. Actinium-225 can be used to treat prostate cancer with metastatic. Single photon emission computed tomography(SPECT) using molybdenum-99/technetium-99m, as well as amyloid-beta positron emission tomography (PET), are used to diagnose Alzheimer's disease, and the Japanese pharmacy company Eisai has also recently started to supply drugs for the treatment of the illness.

In 2010, the Government formed the Japan Nuclear Human Resource Development Network, run by the Japan Atomic Energy Agency and the Japan Atomic Industrial Forum. Universities, institutes and industry and utility-related organizations have joined the network to participate in the education and training programme. The Network has a Steering Committee and a Strategic Working Group with five subcommittees for elementary and secondary education (i.e. high school, junior high school and elementary school), higher education (i.e. university and graduate school), professional-level human resources development (i.e. engineers and industry employees), internationalization of Japanese students and engineers, and overseas education of foreign engineers, officers and researchers.

The Japan–IAEA Nuclear Energy Management School (NEMS) is a highlight of the Network's activities. As you know, the IAEA established NEMS at the Abdus Salam International Centre for Theoretical Physics in Trieste, Italy, in 2010. Japan launched its first School two years later, and Schools have since been held in other countries, including China, the Russian Federation, South Africa and the USA. I have heard that NEMS now has almost 2000 alumni, almost 500 of whom participated in the Japan-IAEA School. This year we will deliver our 12th NEMS, which will include the usual visit to Fukushima Daiichi NPP.

The Ministry of Education, Culture, Sports, Science and Technology established the Advanced Nuclear Education Consortium for the Future Society in October 2021, which brings together universities to share courses and provide opportunities to participate in experiments, exercises, site visits and international study.

Moreover, the JAEC is strengthening nuclear knowledge management by promoting professional engineers status through the Japanese Professional Engineer qualification. The USA, for example, is very advanced; in 2019 it had 890 000 Professional Engineers. The UK has the equivalent status of Chartered Engineer, and France has the State-certified engineering degree (IDPE) issued by the 'grandes écoles' in engineering. Professional mechanical engineers are essential in ensuring regulatory compliance, including new build, design, drawing, analysis and construction in several states of USA. I will elaborate further on the importance of Professional Engineers for nuclear knowledge management tomorrow at the fourth plenary session.

Thank you very much.

Verbatim as delivered.

K. Mrabit
President of PAM Parliamentary Group, House of Councillors, Morocco

Let me start by thanking all the distinguished representatives of Member States attending this very important conference. I think on behalf of all of us, let's sincerely thank the IAEA Secretariat for organizing this important conference and, in particular, the Department of Nuclear Energy and also the division responsible for nuclear knowledge, represented here by the DDG and the Director, for holding this conference several times to organize such conferences, which are very enriching for all of us. And special thanks also to the scientific secretaries who designed, in cooperation with the Programme Committee, this important conference.

Having said that, let me first start with the introduction to put things into perspective and to reach a common understanding, or remind you about some common understanding that we use in Morocco to build our knowledge management and human resource development programmes. I will then emphasize the importance of international and national commitments based on the instruments or conventions that have been ratified by Morocco. For example, a national strategy on education and training for safety, security, and safeguards that we have established in Morocco. This has been supported by capacity building centers that we believe are key in developing, implementing, and sustaining all these efforts. Finally, I will quickly discuss some challenges we have faced and some conclusions.

I think we agree here, and let's remind ourselves that the definition of knowledge management is an integrated, systematic approach to the identification, acquisition, transformation, development, dissemination, use, sharing, and preservation of knowledge relevant to the achievement of specified objectives. This is taken from the IAEA documents. This is based on three important pillars: first, the processes – the procedures that govern all this; and the technology that is necessary to store the information and make it easy for all of us to use it; and most importantly, the people, because you may have the best processes, procedures, and technology, but unless you have an adequate number of well-educated and trained people, you cannot sustain all these efforts. So, these are the three elements on which we in Morocco build our system.

Why does nuclear knowledge management matter in nuclear activities? Like any highly thematic area, the use and continuous improvement of nuclear and radiological technology or applications rely heavily on the identification, accumulation, and innovation of knowledge and experience. This is really key, but it has to be continuously improved. We thank the IAEA for also contributing to this continuous improvement of knowledge management, education, and training in several areas.

As it has already been mentioned, a lot of mature people (I will not say old people like me, but mature people) are retiring, and these people have a good deal of experience and knowledge. This is a real challenge to ensure, retrieve, and keep this knowledge, and to keep nuclear energy and nuclear techniques and applications safe and secure.

The IAEA has been aware of and assigning high priority to knowledge management, education, and training. Let me remind you that many years ago, in 2003, the DG, Mohamed ElBaradei, mentioned that "whether or not nuclear power witnesses an expansion in the coming decades, it is essential that we preserve nuclear scientific and technical components of the safe operation of existing facilities and applications. Effective management of nuclear knowledge should include succession planning for the nuclear workforce, the maintenance of the 'nuclear safety case' for operational reactors, and the retention of nuclear knowledge accumulated over the past six decades."

So, since that time, the Agency has been assigning high priority to this and organizing conferences, programmes, and activities to improve it. What should be done to improve these measures? The aggregation and sharing of knowledge should be promoted, and a more open, knowledge sharing culture should be fostered within organizations. Each organization should play its role. This was key for us in Morocco when we built our system. Starting and implementing knowledge management and sharing in an organization must be done by its staff, by insiders. You may get support from outside, but this should heavily rely on the people within the organizations. A key element for this is international cooperation. I would never repeat it enough what the IAEA has been doing to bring us together to share knowledge and experience on education and training, as well as in technology, safety, and security, – is crucial to all these efforts. Therefore, sustainable education and training, and what the IAEA has been doing, is also key for improving and sustaining all these efforts.

Education and training cannot be dealt with separately. They are part of capacity building, which each country is trying to preserve and improve. Based on the definition of the IAEA, in Morocco, education and training are part of several elements for capacity building. These include human resource development, knowledge management, and knowledge networks. Here, we are also like a network, sharing our knowledge. These are the elements on which we built our system in Morocco, taking these components into account.

The importance of international commitments is key for any country dealing with nuclear energy or nuclear applications. In my country, we have ratified all relevant safety instruments and international conventions on safety, security, emergency planning and response, and other relevant conventions. And I think this is important for each country embarking on nuclear power or expanding it to ratify all these relevant conventions.

Based on this and on the international standards and guidance of the IAEA, as well as best practices, we developed our system in Morocco, taking into account what they have mentioned. This is our system.

At the national level, we have the Head of the Government, and we established a Regulatory Body under the Prime Minister of Morocco. This was not the case before. And let me be frank with you: until 2014, Morocco had two regulatory bodies that dealt with both promotional activities and control activities. This was not fully in line with the international conventions. Therefore, Morocco, with the help of the IAEA, revised its legislation and prepared a law that addresses safety, security, safeguards, and emergency planning and response. This law created an independent regulatory body that reports to the Prime Minister (not to ministries).

This is why, on the left side of the slide, you see the regulatory framework with a regulatory authority – AMSSNuR, the Moroccan Agency for Safety and Security. I had the pleasure of being the first Director General to establish it, and now it's running relatively well. This was a big milestone, a big achievement.

On the right side, in addition to this, we developed policies, and I want to emphasize this. You have the regulatory body, the legislation and the regulations, but you also have policies at the governmental level. We developed policies for nuclear energy, for nuclear safety, for nuclear security, public communication, and human resource development. A specific policy on education and training has been developed at the national level.

The rest of the framework is similar to that of other countries with ministries, followed by operators dealing with nuclear applications, national support organizations, and so on. This is the structure we have in Morocco.

Our system was based on the international standards and guidance, and we took into account the IAEA standards. When it comes to knowledge, education and training. We recognized the importance of the processes to be established. We used not only safety and security guides, but also IAEA nuclear energy documents and the well-known milestone document.

So, we have established the law and, of course, a regulatory body with a clear mission, vision, and values. Furthermore, we developed our policies and, importantly, we created our integrated management system that incorporates all processes, procedures and policies, that also deal with knowledge management, education and training. This is only to show that our integrated management system includes all these elements.

A few words about the strategy on education and training. We established a national strategy for safety, security and safeguards. These are the objectives that were accepted by all partners in Morocco, and these are the main results. The main results, quickly, only to say, that we identified that we need to train more than 14 000 persons in different fields such as medical applications, industry, agriculture, research, and so on. And then, when it comes to implementation, we had more understanding with universities, our technical support center, CNESTEN, and the national radiation protection organization. And, we have signed memorandums of understanding with many organizations and universities within Morocco to implement this. In addition, we had memorandums of understanding with other organizations in different regions and also with the IAEA.

To support all these activities, we created what we call capacity building centers in close cooperation with the IAEA. We established one for emergency planning and response, another for security, and others within the technical support organizations in Morocco.

So here, just a few words about one specific center – the emergency planning and response. I will not cover this in detail, because I see I have 1 minute 50 seconds left. I just want to mention that these centers not only support our national activities, but they are also made available through the IAEA for regional activities in Africa, as well as for international events involving other countries.

What were the challenges we faced? There were several challenges. One of them was to establish and implement a national policy and strategy for education training at the national level. Involving universities and technical support organizations was not easy, but in the end, it was established and approved at the highest level, meaning by the government, and it is now being implemented.

Another challenge was at the regulatory level. We also had to establish and implement our integrated management system, which includes education, training, and knowledge management, and we had to sustain this effort.

In conclusion, the ratification of all relevant nuclear safety and nuclear security instruments is key. As in any country, when you ratify an international instrument, it becomes a law. All these instruments became laws in Morocco, and that was very important in helping us establish our framework. The establishment and implementation of national policies and strategies for education, training, and knowledge management are more effective when there is agreement at the government level. This helps sustain all these efforts.

Capacity building centers are also very important for national, regional, and international knowledge management, as well as education and training. I cannot emphasize enough that the IAEA has been great in supporting all these activities at the national, regional, and international levels. We are thankful for all the support we have received from the IAEA and wish the Secretary to continue and reinforce these activities.

Thank you for your interest. Thank you very much for your attention.

As prepared for delivery.

T. Terentyeva
Deputy Director General, Rosatom, Russian Federation

Inspowered by Atom: Eight Decades of Commitment to People and Planet

1. Introduction

Dear participants, esteemed colleagues, dear Chairs, it is a great honour to stand before you today. I would like to extend a warm welcome to all of you and express my sincere gratitude for this opportunity to share and exchange our ideas and perspectives.

On behalf of the Rosatom State Atomic Corporation, I am privileged to introduce our vision for company growth, centred on the development and empowerment of people.

Our commitment to advance human potential drives every initiative we undertake. Our growth not only benefits the industry but also contributes to the well-being and professional advancement of our workforce and communities.

2. Rosatom at a glance

Today, Rosatom is responsible for producing over 20% of the total power generated in Russia and we stand as a key player in energy balance system in our country. By 2030, share of nuclear in energy mix of Russia will be 25%.

That makes us actively diversify our business to embrace new opportunities and ensure sustainable growth.

Today, Rosatom stands as an employee community of 370,000 people, spread across 460 enterprises.

We operate in 31 nuclear cities in Russia where Rosatom is a key city-forming industry impacting 2.5 million people through our business and social activities.

3. Rosatom being local globally

Our global presence spans 60 countries, encompassing a diverse range of businesses.

We are simultaneously constructing 22 power units in 7 countries. The scale of the projects and their geography makes us to be very focused on the quality of project management and transfer of knowledge from one site to another.

Rosatom ranks in the top three globally in all key areas of the nuclear fuel cycle – uranium reserves, mining, enrichment, and nuclear fuel fabrication.

Our strength lies in offering our partners more than just a nuclear power plant. We build entire industries, complete with scientific research, regulatory frameworks, localized production, and human capacity.

Sustainable development is a key part of our strategy and all operations. We believe that our efforts contribute significantly to the global sustainable development agenda, ensuring a secure and prosperous future for all. In 2020, Rosatom joined the UN Global Compact which underscores our commitment to the UN sustainable goals.

Such a comprehensive approach assures us that all infrastructure and human capacity is in place for the time the new NPP is launched. As a result, our partners get access to a new level of technological sovereignty for their respective countries. Our work is built on the principles of cooperation and respect.

4. Rosatom product portfolio

Our capabilities span from uranium mining and nuclear power plant construction and operation to advancements in nuclear medicine, hydrogen energy, energy storage, electric vehicles, wind energy, and other environmental solutions.

Rosatom produces over 70% of the world's range of medical isotopes. We export isotope products to more than 50 countries.

Additionally, we are exploring the commercial potential in Arctic Zone in the Northern Sea Route through our fleet of nuclear icebreakers.

In promoting excellence for such a global and diversified business, the core task and challenge for us is to enhance employee sustainability, ensuring both business viability and long-term results.

5. Keeping the pace

Here you can see the outline of benchmark events in our history which had a major impact on how the HR systems evolved in order to keep pace with requirements of our business.

Next year, Rosatom is celebrating its 80th anniversary. Just few days ago we celebrated a very important milestone for Russian and global nuclear industry. On June 26th, 1954 the world's first nuclear power plant was launched in the small city of Obninsk, marking the birth of commercial power generation.

In the 2000s, we experienced a significant breakthrough known as the atomic energy renaissance when Rosatom was actively adding new NPPs to energy system of Russia.

Since 2010, we have undertaken large-scale construction of NPPs both in Russia and internationally. We have also added wind generation and nuclear solutions for the Arctic.

Since 2020, we entered a new decade which is marked by technology breakthroughs. Our product portfolio has significantly diversified to include Small Modular Reactors (SMRs), floating NPPs, Generation IV technologies to solve spent fuel problem, etc.

Whatever business we are developing, our firm belief is that we are responsible for the communities we provide services to. Our corporate responsibility programmes cover not just cities where Rosatom is operating but expand to national and global initiatives in education, art and culture which reach millions of young people and people in local communities.

6. HR Outlook

Our employees are the backbone of our success. Their commitment and expertise enable us to continue leading the way in technological advancements and sustainable energy solutions.

We form a very diverse and inclusive workforce and spaces that welcome individuals of all ages, mindsets, genders, skills, and abilities.

We have a balanced age structure within our workforce, with women making up 32% of the total. We actively encourage female participation through leadership development programmes and outreaching to universities to attract girls into nuclear industry.

Over 5,000 of our staff have doctorate degrees, over 15% are senior-age experts which proves our position as a knowledge company.

We are proud that the results of our work are recognized at the highest level by professional HR community and our employees.

We have been a leader in the national best employer ranking for over 5 years while competing to companies from other industries.

But what is more important for us is the engagement rate of our employees which is comparable to the most attractive employer brands in IT, banking, and consumer goods.

7. Top HR Priorities

People are at the centre of any transformation. We have weaved the principle of human centricity into the Rosatom strategy. We aim to create a corporate environment where every employee can unleash their potential. Additionally, we are committed to supporting our employees' well-being, offering robust happy retirement programmes to ensure a fulfilling post-career life.

There are three pillars that are currently being developed within our HR strategy to achieve Rosatom's Vision 2030. They are Values and Culture, Leadership and Management and finally – Trainings and Development.

Each pillar is a system integrating advanced HR solutions to create opportunities for personal development of each employee, fostering comfortable teamwork environments, well-being of our employees and transparent and fast management structure.

8. Capacity Building ECOSYSTEM

Our company offers a life-long path for our employees, beginning with schools through university programmes.

We provide advanced opportunities to improve and grow within our organization. That's why we created learning ecosystem, including continuous learning, upskilling programmes, and career advancement opportunities. They include two corporate universities that provide over 600 training programmes and over 3 000 online courses, with Science and Innovation being in the forefront of our training portfolio.

Vocational training is a big area of our focus as workers form one third of our employee pool. We have developed AtomSkills standard which is a unified methodology for training of engineers and blue collars. It includes a set of methodological materials, community of practice and annual AtomSkills championship that allows our workers and engineers to test their practical skills and to learn from each other.

Through this holistic approach, we foster a dedicated and skilled workforce, empowering our employees to achieve personal and professional excellence at every stage of their career.

9. Youth outreach

We place special emphasis on shaping the future human capacity for our business. Currently, one third of our employees are under 35 years old, and they are deeply involved and engaged in our corporate life. Our corporate students' union and various activities ensure a smooth integration of young people into corporate life.

We annually recruit the top 3,000 university graduates and have a strong employer brand for young engineers.

We are committed to working with youth, collaborating with school communities to inspire children to pursue STEM professions.

Our academic network includes the 20 best Russian engineering schools, including the leading nuclear research university MEPhI (Moscow Institute of Engineering and Physics). Currently over 2,000 international students are studying at Rosatom's partner universities in Russia to train highly skilled professionals for NPPs in newcomer countries.

10. HR-innovation

Technologies like AI are transforming the way we live, communicate and work and bring huge benefits.

We are in the process of developing exciting new HR digital products. They put the user experience in the centre through personalised services. These include social network, digital solutions at all steps of HR cycle starting from recruitment to retention, etc.

HR-analytics is an area we are heavily investing in to keep pace with the development of our business. I am proud to say that currently over 250 companies are participating in our research programme called Human-Centricity Alliance. Together we are conducting HR studies and creating solutions to help businesses to adjust to challenges of labour market.

Thank you for your attention!

I cordially invite you to learn more about our HR and knowledge management activity at our booth and we will be happy to meet with you at our side event on 3 July.

We are just at the beginning of the very important journey aimed at creating innovative shared responses for nuclear industry to triple power generation capacity over the next three decades! I am happy to see that global HR community and our academic partners are taking action to be part of this joint effort!

I believe that the key to success is in fostering greater cooperation and working together on joint efforts. I wish all of us a very productive conference!

As prepared for delivery.

R. Griljov
Managing Director, Paks II. Ltd., Hungary

As for the current status of the Paks II. project, it is progressing well, we are at the phase before the first concrete. Production of the first large nuclear island system, the core catcher is completed (it has been delivered for Unit 1 of the Paks II. nuclear power station in August 2024 – Ed.) and the reactor pressure vessel is also under manufacturing. In June 2024 the project received exemption from EU sanctions and European companies will no longer need the permission of national authorities to participate in the project.

Our country has various institutions linked to the nuclear industry with the need of the right experts, and we also do our best to ensure that the Paks II. nuclear power plant can be operated properly.

It is important to add that according to a 2017 European Commission decision, Paks II. is functionally and legally separated from the operator of the Paks Nuclear Power Plant (the incumbent MVM Group) and any of its successors or other state-owned energy companies to avoid market concentration. The decision of the European Commission needs to be taken into account during the education of the personnel.

Our human resource programme is based on the students. Through the scholarship, internship and fresh graduate programme we try to make attractive this field for the students who finished their education. With this programme we want to transfer the knowledge from the experts to the next generation.

How we try to prepare the students for this industry. First step is ESZI Institute for Energy Vocational Training, a secondary school where the students first can meet the nuclear industry during the education programme financed jointly by Paks I. and Paks II. The school is seeking opportunities of international cooperation with other similar institutions. We are cooperating with universities as well. Paks II. Academy, in operation since 2019, provides a great opportunity to offer basic nuclear knowledge to students. We have contracts with six universities with technical faculties in Hungary for nuclear power plant operation engineer and technician courses. Our new focus is to upgrade this project and offer a BSc education programme in the region of Paks. At first, electrical engineering training will be launched. If successful, other trainings might be launched later.

Paks II. Ltd. considers corporate knowledge management too. Our company established a huge online platform, a database where the employees can find conference materials and publications related nuclear industry. Our experts have the opportunity to give monthly presentations on different topics and we have also online courses and tests for the colleagues (nuclear safety culture, HSE, ESG, etc.).

Besides, Paks II. has an educational contract for common training, special nuclear power plant training and practical knowledge acquisition with Paks I.

For us, it is important to exchange experience with countries that operate or are constructing nuclear power plants with similar technology, we regularly organize workshops with international partners in cooperation with WANO.

As prepared for delivery.

K. Pringle
Director of Human Capacity Building,
King Abdullah City for Atomic and Renewable Energy (K.A. CARE), Saudi Arabia

Excellencies, distinguished delegates, ladies and gentlemen,

On behalf of the Kingdom of Saudi Arabia, I extend our sincere gratitude to the International Atomic Energy Agency (IAEA) for organizing this vital conference on 'Nuclear Knowledge Management and Human Resource Development'. It is an honour to address such a distinguished gathering of experts and policymakers committed to the peaceful use of nuclear energy.

We recognize the critical role of effective knowledge management and human resources practices in ensuring the successful and sustainable operation of nuclear facilities. Addressing challenges such as knowledge transfer, and safety culture is essential for fostering a skilled and adaptable workforce.

By implementing knowledge sharing policies, continuous training programmes, and promoting a culture of learning and innovation, professionals can better navigate the complex nuclear landscape. The IAEA plays a key role in facilitating global collaborations, sharing best practices, and supporting nuclear organizations in enhancing operational efficiency, safety performance, and overall sustainability.

Collaborative efforts and knowledge exchange are essential in developing practical solutions to support the development of a knowledgeable workforce for the safe and sustainable future of nuclear power programmes worldwide.

The Kingdom is committed to its national policy for atomic energy, which ensures the highest standards of transparency and reliability, and the highest levels of safety. The Kingdom recognizes the positive contribution of nuclear energy to energy security, as well as its socioeconomic benefits.

Therefore, the Kingdom is working to develop nuclear energy across various fields through close cooperation with the Agency, which will help meet the sustainable development goals as outlined in the Saudi Vision 2030, according to both local requirements and international commitments.

Additionally, Saudi Arabia recognizes the importance of 'nuclear knowledge management and human resource development' as crucial pillars for the sustainable and safe deployment of nuclear technology. As a nation committed to diversifying its energy mix, we see nuclear energy as a key component of our future energy strategy.

To this end, Saudi Arabia has prioritized early on and embarked on a comprehensive nuclear energy programme aimed at developing a robust and knowledgeable workforce. Our efforts are focused on several key areas:

1. Education and Training: We are committed to supporting educational institutions to cultivate a new generation of nuclear scientists, engineers, and technicians. Partnerships with leading international universities and nuclear research centres are being strengthened to ensure our workforce is equipped with the latest knowledge and best practices.

2. Knowledge Management Systems: Recognizing the value of effective knowledge management, Saudi Arabia is working closely with the IAEA to ensure awareness and adoption of best practice of implementing advanced systems to capture, store, and disseminate nuclear knowledge. These systems

will be designed to ensure continuity, avoid knowledge loss, and facilitate the sharing of expertise across the sector.

3. International Cooperation: The Kingdom is a key proponent of international collaboration and is actively engaging with the IAEA and other global partners to exchange knowledge and experiences. Saudi Arabia is committed to contributing to the global nuclear knowledge base and learning from the experiences of other nations.

4. Human Capacity Building Strategy: The Kingdom has made significant strides in the development of its Atomic Energy Human Capacity Building Strategy. We recognize the critical importance of developing a robust and sustainable human resource framework to support our peaceful nuclear energy ambitions.

Our Atomic Energy Human Capacity Building Strategy is designed to prepare and qualify a sustainable national workforce that will contribute to the energy transformation of the Kingdom. The four essential components of our strategy were to first develop the nuclear energy human capacity building workforce plan and roadmap. Second, we established strategically important specialized educational and training programmes through partnerships. Third, we initiated employment and training mechanisms through industry experience and summer programmes, and lastly, we designed a Digital Platform to enable, support, and facilitate the delivery of the Human Capital Plan.

Saudi Arabia is dedicated to playing a proactive role in the global nuclear community. We firmly believe that through strategic investments in knowledge management and human resource development, we can achieve our vision of a sustainable and secure nuclear energy programme.

In conclusion, we reiterate our commitment to the IAEA's mission and to fostering a culture of safety, and excellence in nuclear knowledge management and human resource development. Together, we can ensure that nuclear energy continues to contribute positively to global energy needs and sustainable development. And once again, thank you, esteemed guests and participants. I look forward to a successful outcome from this conference.

Verbatim as delivered.

Presentation: Kingdom of Saudi Arabia Human Capability Build-Up Strategy for the Atomic Energy Sector

We started with the national commitment for the programme, which is to build two nuclear power plants. Depending on the capacity or the technology, this is what we had in mind. With that as our target, we then needed to focus on building the objectives of the project, which were to ensure that we're building the Kingdom's human capabilities and culture to support operational readiness and make sure that we have them ready in time for the project.

We had to formalize and operationalize a long term customized human capacity building strategy. To do that, we had to first determine what the current and the future needs for the sector were. This was very important. We also wanted to make sure we had operational readiness from a local perspective. And we wanted to make sure that within our own department, we were building a sustainable workforce that could carry out this national strategy.

Overall, we were trying to overcome four main challenges and I'm sure, these are challenges that many of you have faced when you wanted to design your strategies.

The first was looking at the workforce requirements. Did we have clear qualifications? Did we understand what the different jobs descriptions were, and what the different competencies and skills were required? So, we needed to make sure that we identified those, and, of course, the IAEA provides great guidance on that. If we didn't do that, it could risk the limited Saudi resources being employed, if we didn't have the right skill sets identified.

The second is the sector attractiveness. This is another issue about making sure that the sector is competitive vis-à-vis other sectors. It is important to ensure that we are making it an attractive option for the future. In Saudi Arabia, we have a lot of young people under the age of 30, so we have a huge opportunity here to really attract talent and build them with the right capabilities.

Also, we had to look at the training and development infrastructure. Do we have the right equipment, the right kind of faculty members, the right kind of curricula and programmes in place? And if not, what do we have to do to ensure that we're building sustainable development within the Kingdom, without relying on organizations abroad? And we could do that with partners.

And then, the workforce retention. As the Director General mentioned in the beginning, the workforce retention is important. And this is also something that we wanted to factor in early on, from the beginning of the strategy.

Here, we looked at what our guiding principles were. We had to define, for the framework, what the required capabilities and capacities were that we needed over time. Second, we wanted to make sure that we're looking at attraction – how are we going to inform, educate, and attract the right talent? But also, involving and training with the right partners, both international and local. And then, how do we make sure that we're securing on time the sustainable workforce that's required? All of that is in collaboration with all the different government stakeholders, educational stakeholders, and industry.

Moving on to our framework, we built a vision and a mission. It was sort of what this human capacity building strategy is all about and how we should go about it? Then, we defined the what are the key questions: What is the supply that we need from the market? What are the job clusters that we require? What is the gap that needs to be filled? So, looking at defining that, what is the challenge?

Then we looked at attraction. What training is needed? What are the target groups that we need to focus on? What are the key messages? How do we compete? In the Kingdom right now, we have Vision 2030, a major national transformation programme. There are many opportunities in other very attractive sectors that Generation Z is really interested in. So, how do we make our sector equally attractive? How do we also make it attractive to women? There are more and more gateways opening for women in engineering in the Kingdom now. So, we're really trying to make sure that we have a clear path for that.

Involving and training. What are the right programmes we need to roll out, and with which partners? What is the right infrastructure and the right channels to execute the programmes?

And then, embarking and retaining. How should employees be onboarded into these sectors? How should we also ensure that we have the right methods for professional development? So, a lot of these aspects had to be considered as we built the strategy. And underpinning it, as the Director General said, is funding. How do we pay for all of this? How do we develop a plan with the government to outline our requirements and ensure we have the necessary funding every single year? And then, governance – making sure we have good alignment across all these different sectors.

Looking at the key themes, we had to ensure a good workforce ramp-up, making sure that we had an adaptable model for NEPIO, owner and operator – who is thinking about all of those organizations. We also wanted to make sure that we are enhancing sector attractiveness to attract the right talent, as well as developing the right partnerships. Then, the last one around retention. So, we had four key themes that we were trying to fill with this strategy.

We had 14 projects that we identified under each one of those. So, you can see for each project, we had a number of different initiatives that we were launching, whether it was raising awareness or creating mechanisms for resource selection. We had identified very specific projects that needed to be completed for each theme of the strategy.

Here, you're looking at the capacity building, which to me is one of the most important elements. Like, you can attract talent, but if you don't build the right programme using a standardized approach to training and ensuring you have a common outcome, you're not going to get there. How do you build capacity in universities? When do you start building the technical vocational elements? When do you start building the plant? There are so many things to think about. So, having a really targeted plan to help us build that roadmap was very important.

We are the national orchestrator in Saudi Arabia for human capacity building across all energy pillars. Nuclear is only one of them. We started off by building this strategy for nuclear, and based on that, now our Minister of Energy has asked us to look at a cohesive approach, breaking down silos and building capacities across energy sectors as a whole. So, maybe we can get some synergies there: we can build more attraction to energy as a whole and then benefit from that when the nuclear programme moves forward. It's really interesting, the position we're in right now, where we're thinking not just within our own nuclear silo, but more broadly across energy in general. There are a lot of commonalities, and I think you can find value in that.

So, overall, looking at the different phases, we began first with the first phase of making sure we're putting all of the foundations in place. The second phase is rolling out the programme and adjusting. You know, you see we have some programmes that we've launched. Then, the operational phase, when we're moving into the actual building and construction of the power plant, making sure that we're operational and ready on time.

Partnerships that we've launched. We have two for nuclear. We have two universities in the Kingdom that are working with us. The first is King Abdulaziz University. They've had, for decades, the only nuclear bachelor programme in the Kingdom. So, we're making sure that that programme will remain sustainable. Females will be joining the program next year, so we'll have both male and female graduates from this nuclear bachelor programme. With another strong university, the strongest for engineering in the Kingdom, is King Fahd University for Petroleum and Minerals. They launched first a nuclear concentration programme, attracting students who are bachelor students in chemical and mechanical engineering, and offering them a concentration in nuclear. This has proved to be very successful.

We've had summer programmes where we've sent the students abroad. We sent them in the 1st year to France with the cooperation of I2EN, which was a very successful programme for 8 weeks. In the 2nd year, we sent the students to Slovenia and Switzerland, which, again, with the support of PSI, provided a really outstanding programme for our students. And I think next week, we're sending our students this year to Korea. So, we're really looking today and during this conference to find more partnerships where we can build collaboration and build partnerships with countries and organizations that can help build our future talent.

In Saudi Arabia, we don't have a mature industry. We need to send our students to countries that do, so they can understand the whole value chain and supply chain of nuclear and how it works. These three experiences that we have provided for our students have been really outstanding.

And this is an overall look at how we, as K.A.CARE – the organization I work for – are positioned in the middle. We play this role as the national orchestrator and pacesetter for the energy sector. So, you have the education and training, and policy making role. You have also the energy policymaking, with the Ministry of Energy having a role. You also have the education providers, both international and local, and then you have the employers. K.A.CARE is at the centre and serves as this national orchestrator.

And lastly, just some highlights of the journeys I mentioned, where we sent our students. We also had the honour last September of our Minister of Energy, His Royal Highness Prince Abdulaziz bin Salman Al Saud, signing for the first time the JPO – the Junior Professional Officer Programme – with the IAEA. We're very excited to be one of the Member States participating in this programme. We are currently underway in attracting different candidates to the programme, and hopefully, we will have a long list of young professionals joining.

So, thank you very much for your time today. I'm really welcome to talk to all of you and learn more about what each of you are doing in your countries and how we could potentially collaborate and work together.

Thank you so much.

As prepared for delivery.

E. Pule
Conference President and HR Executive Eskom Holdings SOC Ltd, South Africa

Good morning, ladies and gentlemen,

As introduced by the DG earlier, my name is Elsie Pule, President of the IAEA Conference. I am from South Africa, a member state of the IAEA since the agency's inception in 1957 and will also be sharing with you our countries view on Nuclear Knowledge Management and Human Resources Development in the Nuclear field.

South Africa and the IAEA

Ladies and gentlemen, we all may be aware that last year, the IAEA celebrated 70 years since the Atoms for peace speech was given by US President Dwight D. Eisenhower to the United Nations. The Agencies statute was approved in 1956 by the Conference on the Statute of the IAEA held at United Nations Headquarters, New York; and entered into force on 29 July 1957. The Headquarters of the Agency are situated here in Vienna. Its principal objective is "to accelerate and enlarge the contribution of atomic energy to peace, health and prosperity throughout the world''.

South Africa became a member state of the International Atomic Energy Agency (IAEA) in 1957. And in 1959, the South African government approved the creation of a domestic nuclear industry and planning began the following year on building a research reactor, in cooperation with the US Atoms for Peace programme.

The Atomic Energy Board and Atomic Energy Cooperation Evolution

With the Atomic Energy Board's formation from the Atomic Energy Act in 1948 and its transition to the Atomic Energy Corporation (AEC) between 1957 and 1961, we have seen some operational changes. In 1970, South Africa embarked on an extensive nuclear fuel cycle programme, as well as the development of a nuclear weapons capability in 1988.

Subsequently, 1993, South Africa voluntarily announced the dismantling of the nuclear weapons programme and in 1995-96, the IAEA confirmed that it was satisfied that all materials accounted for the nuclear weapons programmes were terminated and dismantled.

South Africa's Nuclear Activities and Future Nuclear Prospects

Ladies and gentlemen, previously, I gloated that South Africa operated the only nuclear power plant in the country and on the continent, but this won't be for long.

Although no new nuclear capacity is planned before 2030, the IRP2019 and IRP 23/24 report indicates that consideration needs to be given to developing the future nuclear programme. The IRP recognises that SMRs have a strong potential for adoption in the country.

It is encouraging that in no time we will see another nuclear project on our continent, with the construction in partnership with Rosatom of El Dabaa Nuclear Plant in Egypt to increase its base load and a nuclear footprint in Africa.

Both the LTO and the New Build necessitate effective NKM and HRD to ensure safe operations, if NKM and HRD is neglected in these two major projects, we may face catastrophic events.

The Case for Extending the Life of Koeberg Nuclear Power Plant in South Africa

Koeberg was built in the 1980s with the first unit (unit 1) commissioned in 1984 and unit 2 in 1985, the station was built with a lifespan of 40 years. This means that this year and next year (2025) we should be decommissioning the stations as we have reached the 40-year mark for unit 1 and 2 respectively.

As a country the focus on nuclear operations is still strong and the case for LTO is viable. Why am I saying that:

- nuclear power generation provides a stable supply of electricity during the Just Energy Transition. The Just Energy Transition incorporates power from renewable energy sources into the national grid but natural sources such as a sunlight for Solar PV and Wind are neither reliable nor are they fully predictable. Nuclear provides that stability to balance the grid;
- Nuclear power generation has superior plant performance: in the context of South Africa (and I can speak for most of the African content), we have had unstable power due to various reasons. Extending Nuclear is thus a natural choice to make when considering the plant performance compared to the rest of the coal fleet: on average Koeberg has higher Energy Availability Factor (EAF) than the rest;
- The plant location was well selected, taking into account various factors for safety; Koeberg was designed with operatorial safety in mind: it is built on bed rock and can withstand severe earthquakes comparable to the Tulbagh earthquake of 1969 (that measured 6.3 on the Richter scale.) If anything, one would want to run this station longer for its structural resilience;
- A cost benefit analysis was done and while the costs of capital are high in the case of nuclear, the operating costs are substantially lower, even lower than that of renewables;
- Finally, nuclear energy has a lower environmental footprint than conventional fossil fuels which are largely coal fired power stations that we have in South Africa.

As mentioned earlier, we anticipate that the nuclear component of our energy mix by 2030 will remain the same or increase.

In Conclusion

There is a unique Human Resources Management opportunity for Nuclear Energy in the South African context across the entire Human Resources value chain namely: workforce planning, inducting, and training and development, remuneration and benefits, DEI&B, employee engagement and change management and well as attrition management.

Let's explore a few of these:

- South Africa has a legacy of racial discrimination, this forced us to employ locals as far as possible. Here I propose a human resource management plan that is planned at a national level (in fact the IAEA suggests 10 – 15-time horizon);
- Collaborative efforts for capacity building between organisations and countries. I am happy to announce a ground-breaking initiative we have recently signed an MoU with ROSATOM that includes various programmes in nuclear including some on gender equality, because the nuclear field is still highly male dominated;
- Building our own timber. Talent is mobile therefore the solution is for organisations to continuously build and sustain their skills pipeline for example, in 2017 we ran the "Nuclear100" programme, and I am happy to share that 40 of these 100 learner recruits were female.

One of the most long-standing frameworks for knowledge management is that of people, processes and tools. This knowledge framework is applied to all fields (not only nuclear) and by many industries and sectors (including management consultancy houses). I have listed possible solutions in each of these. I hope these are relevant as we discuss collaboration opportunities in this conference over the next week.

That's it from me, I hope you have found this valuable, and I look forward to engaging with you outside the session or in-between the breaks that is where I will take questions and we can engage further so please do come through to my booth to discuss and ideate any problems we can jointly resolve.

Thank you.

As prepared for delivery.

A. Al Mur

Director of Human Resources, Federal Authority for Nuclear Regulation, United Arab Emirates

Distinguished guests, ladies and gentlemen,

Good morning,

At the outset, I would like to state that there are about 440 nuclear power plants operating in almost 32 countries, generating about 10% of the world's electricity. In addition, there are some 60 reactors under construction and a further 110 are planned globally. This fact underscores that the nuclear sector relies heavily on a specialized and knowledge-intensive manpower for its operations, safety and future sustainability. Organizations are depending on the investment of the human capital to be enabled to sustain its business and fulfil its mandates.

The United Arab Emirates has always recognized the critical importance of the investment in the tools that would safeguard the capturing of the operating experience, marinating the knowledge for current and future use, and the continuous management of the development of the human capital and human resource development in ensuring the safe, efficient, and sustainable use of nuclear technology.

The UAE underscored the importance of building the capacity of its nationals from day one of building its peaceful nuclear energy programme. It has developed and implemented various capacity building programmes to ensure Emiratis are equipped with the needed talents and competences necessary to ensure the safe operation of its nuclear power plant. We are proud to say that the UAE has almost 20,000 workers in the nuclear and radiation sectors, which reflects the massive efforts undertaken by the country over the past years. The UAE Nuclear Energy Programme has become a role model for Member States in building and operating its nuclear power plants adhering to IAEA and international safety standards.

Nuclear knowledge management is important in maintaining and transferring the knowledge and the skills that supports the safe and effective use of nuclear technology. As the nuclear industry evolves, the need to manage knowledge across generations becomes vital. This involves not only capturing and preserving the knowledge but also ensuring the tacit knowledge held by experienced professionals is effectively transferred to the next generation.

In the UAE, we have developed an integrated knowledge management framework that includes systematic documentation, advanced IT solutions, and robust training programmes. Our efforts align with the IAEA's guidelines to ensure that nuclear knowledge is not only preserved but also continuously enhanced and shared.

Equally important is the development of human resources. The UAE places significant emphasis on building a skilled and competent workforce capable of operating and regulating nuclear facilities with the highest standards of safety and security. Our human resources strategy encompasses education, training, leadership and continuous professional development.

The UAE has academic programmes that focus on building national capacity in the nuclear industry. The UAE invests in various youth programmes to support knowledge management and professional development, recognizing that 44% of the Federal Authority for the federal authority for Nuclear Regulation workforce is comprised of youth, with gender balance nearly at 50%, and technical roles filled by women at approximately 37%.

There is no doubt that the nuclear sector faces a number of challenges: one of the challenges is the advancement of new technologies, which requires continuous upskilling, and reskilling of the employees. The UAE has addressed this by investing heavily in education and training programmes. Initiatives such as the UAE's National Innovation Strategy focus on providing the necessary resources for training and development, ensuring that the workforce remains competitive and capable of meeting evolving industry demands.

'Employee Mental and Emotional Health': Addressing mental health issues has become increasingly important. The UAE has implemented several initiatives to promote employee well-being. These include creating supportive work environments, offering mental health resources, and promoting work-life balance. Programmes like the National Programme for Happiness and Wellbeing demonstrate the UAE's commitment to enhancing the mental health of its workforce.

'Diversity and Inclusion': The UAE has made significant strides in this area by implementing inclusive HR practices and promoting diversity in recruitment and advancement. The UAE Gender Balance Council is one such initiative that aims to ensure equal opportunities for all genders in the workplace, fostering a culture of inclusivity.

'Remote and Hybrid Work Models': The UAE has embraced this change by developing robust digital infrastructures and remote working policies to ensure productivity, engagement, and compliance with labour laws. This has enabled a seamless transition to new working models, maintaining efficiency and workforce satisfaction.

'Leadership Development and Succession Planning': Developing effective leadership pipelines and succession plans is essential to ensure the seamless transition of talent into critical roles, sustaining long-term organizational success. The UAE focuses on nurturing leadership through targeted training programmes and strategic succession planning, ensuring a steady supply of capable leaders for the future.

The UAE is committed to transferring its operating experience model to nuclear newcomer countries with a special focus on knowledge management and human capital development. We recognize that the challenges we face are complex, but we are confident that through collaboration, innovation, and dedication, we can overcome these challenges and build a sustainable and prosperous future for the nuclear sector.

Thank you.

Verbatim as delivered.

A. Duncan
Deputy Assistant Secretary for International Nuclear Energy Policy and Cooperation
Department of Energy, USA

I am honored and pleased to be here with you today. I'd like to thank the IAEA and the Scientific Secretariat for organizing such an important event. And as someone who organizes conferences, I think we can't thank you all enough because you all are the real magic in what I know will be a successful discussion this week.

So, I want to start, unlike my good friend, Julián Gadano, who quoted himself. I think all of you need to go back and write some remarks where you quote yourself because that takes your remarks to the next level. But I want to start by referring to something that Director General Grossi said this morning. He said it's an interesting moment full of problems. And oftentimes, when people talk about human resource development and knowledge management, they look at what they're doing. It sounds great. And in fact, sometimes it sounds easy. But I want to take a different spin on both the US approach as well as what we're doing internationally and share with you some of those problems, challenges, and opportunities that we've identified in these areas.

I'll start with human resource development. If you can imagine, in a non-state-owned environment with many agencies, ministries – whatever you might call them in your country –each of them have to manage their resource development needs, as well as, industry, which is non-state-owned, non-governmental organizations, which are non-state-owned, academia, national laboratories, who evolved in the research and development – all of those various entities have to manage human resource development.

And so, trying to say there is a strategy for all of that can be very difficult. It is like deciding a menu for a very important holiday in your family, and we know how difficult that is. But I think it's critically important that we look at some of these challenges so that, as we move forward, we are able to help these emerging countries who will have to do the same thing by presenting a picture of the things that they need to think about. From the U.S. perspective, as someone coming from the U.S. Department of Energy, we have had some very awakening information come to us. And one of the biggest challenges we face is from our seasoned employees. I like to call myself seasoned – some people call them senior, some call them retiring or older people – but seasoned sounds a little bit interesting and spicy, doesn't it?

So, from our seasoned professionals and their pending retirements, that leaves a huge gap. Most people have looked at this gap in terms of the technical information that gets lost. But I submit to you, there's an entire culture of change that has to happen. It's not just about new workforce entrants stepping into that technical space – those are leaders. Those are policy influencers. Those are the people who would have to make decisions, should there be any type of implications in the nuclear reactor space. Those are the people who are gonna decide what the curriculum is. Those are the people who are going to decide how to integrate the innovation. And the list goes on and on and on. An interesting moment full of problems.

And when we think about how we manage that, we have to also look at this new generation of people. And I don't care where in the world you are – they're very different. They are cut from a different cloth. I haven't figured out where that cloth comes from just yet, but they have opinions. They have rights. They want to make change. They want to be in charge tomorrow. And we have to look at how we attract them because, unlike a lot of us who grew up with bills to pay, money is not necessarily a motivator.

We've had to study that intergenerational dynamic for two to three decades now, even since our seasoned selves realized that not everybody is going to be as disciplined and rigid as us – drawing on those cultural and historical ties that we have from our family. This group is different. And we have to integrate that into our thinking as we look at how we bring them into our workforce.

Because they want to work on impactful things. They want to work on exciting things. They want to actually be able to do something. And so, when we think about how we attract them, we have to think about that in a very creative way. We also have to empower the voice that they have, the drive that they have, the passion that they have. And I would say – we need them. For those of you in the audience who consider yourselves young, that's not my place to put that label on you – but we need you. And when we look at, again, bringing folks in, we have to look at the folks who are doing the welcoming. Are we really welcoming them? Are we really considering what they believe to be important? Are we really looking at their ideas?

And so we've thought a lot about this in the United States because this is no easy task. And I can't say that we have all the solutions, but at least we understand that there is a challenge for us to overcome.

We've worked in collaboration with government agencies, industry, NGOs, colleges and universities – all of the stakeholders involved in what will be a workforce that has pledged to triple nuclear, that has pledged to triple renewables (which will be integrated with nuclear), and that has pledged to really grow our domestic capacity. And, by the way, we also help our friends and neighbours around the globe do the same. And this is a time filled with problems, but the most exciting time in nuclear that I've ever seen. And I've spent more than and I'm not afraid to say my age. I've spent three and a half decades in this work, and this is the most exciting time.

What we do know is that the technology is going to move forward. We've all seen the commitments made at COP with 20+ countries. We've seen the additional commitments around renewables, as I've mentioned, at COP as well. But where is that workforce going to come from, and are we paying attention? I'm glad to see that there are some African countries in the room today, because while you've heard that Africa is without electricity in about 60% of the continent. I wonder if we're all aware that Africa's youth are going to make up more than 40% of the globe's workforce. So, when we think about what we're doing, are we really factoring in all of these things?

From the United States' perspective, we've prioritized Africa to really support the needs there. Because human resource development starts way, way, way, way, way before somebody's going to retire. And in fact, we've reached down even to six-year-olds to start those conversations. Because while there might be a six-year-old who may want to be a physicist, an astrophysicist, a nuclear engineer. I don't hear many of them saying that they wanna be a welder. And yet there are countries that pay $50K for two weeks' worth of welding, because that skill set simply doesn't exist.

I don't know about you all, but I don't make $50K in my position. But I think it's critically important that we look at the breadth, and we look at the young people that we want to bring in. Because we not only have a responsibility nationally, but internationally we have a responsibility because we all have the same shared challenge. And there simply will not be enough of human resources to go around for all of us. So, the time to invest in our friends and partners around the world to solve these shared challenges has already started. I can't even say the time is now. It has started. And for some of us, we're still behind because we're asking stupid questions like why Africa? Are they ready? They are. They've got so much talent there. And we're very pleased to partner with the IAEA on its Lise Meitner programme, which has had two cohorts in the United States and one in the Republic of Korea to welcome African, European, Asian students from around the globe. So that we can start to work with these populations. And Lise Meitner is a mid-career programme. And I say this to you, so that you're clear that it's not just these people who are starting out who need this human resource development. It is needed throughout the mid-career and beyond. Because when we talk about retention, we have to make sure that we keep these people trained. Not just keep them in their respective workplaces, but keep them trained. Because now we'll have to take that dimension of having people understand that they'll have to be consistently trained. And we've started to look at that in the United States because there have been many cases where we think we're going to lose some resources, but there could be consulting, training, mentoring, and other opportunities to use our experienced workforces. And we've started to help other countries do the same.

You've got this expertise; you can't just let it walk out the door. There is a way to continue to work with that. And the United States has started to look at those issues too and work with countries, specifically in Europe, to help with that. Because we don't want to lose the knowledge anywhere. It's not just a national issue. So, we have to make sure that we continue to work towards it.

And quickly, as I move to think about some of the strategies we have in the United States in terms of knowledge management. Someone mentioned succession planning. And that starts again well before people want to walk out the door. Mentoring is a big part of that, and we all have a responsibility to mentor others. I was so proud that DG Grossi held this year on International Women's Day an event which pulled all of the women from the Marie Curie Fellowship programme, as well as Lise Meitner programme. We had an auditorium full of women from mid- and early-career stages, and some even a little bit past their career stages. There was that opportunity where we were planting seeds, so that those people go out, replicate, scale and expand what they are doing, ensuring they continue this cycle. And I'm very pleased that the United States will support the work on leadership and resilience that my friend Lisa Lande is doing, so we look at this very critical skill set of leadership. Because without it, all of the technical innovation will have a huge gap of folks to take us forward.

So, that I hope that over these next few days, we have an opportunity to talk about things outside of the technological advances. Recognizing that we're not only looking at a workforce, we're not only looking at knowledge management, but we really have to look at a new culture, a new way of doing business. And the fact that these people, whom we want to welcome with open arms to take us into the next generation of advanced reactor technology, need more than we've ever delivered before. I know we have the right group of people in this room to have those discussions.

I thank you for your time and attention.

4. WOMEN IN NUCLEAR IAEA: STRATEGIES FOR RETENTION AND CAREER DEVELOPMENT

Panel discussion was moderated by J. Donner, President of WiN IAEA.

Opening statemen was delivered by M. Chudakov, Deputy Director General and Director of the Department of Nuclear Energy, IAEA. He highlighted the significant strides made by the IAEA towards gender parity, emphasizing IAEA initiatives like the Marie Skłodowska-Curie Fellowship Programme and the Lise Meitner Programme, which have supported women's education and professional development in nuclear fields globally. He noted the Agency's progress under Director General R.M. Grossi, including increasing the number of women in professional roles within the nuclear sector.

Panellists:

E. Pule, HR Executive, Eskom Holdings SOC Ltd, South Africa

E. Pule shared Eskom's efforts in promoting gender equality, emphasizing the need for a conducive work environment. She described Eskom's multi-faceted approach to gender inclusion through five key streams: fostering women in leadership, enhancing women's roles in technical fields, implementing gender-friendly policies, increasing women's participation in nuclear roles, and building partnerships for capacity development. These efforts involved creating supportive environments, addressing patriarchal perceptions, and ensuring women were intentionally prepared for impactful roles. Partnering with universities, it developed programmes to train women in power generation and distribution. The result was a significant increase in women's representation in leadership and technical positions, a reduction in the gender pay gap, and a shift towards evaluating employees based on contributions rather than gender.

K. Pringle, Director of Human Capacity Building, K.A. CARE, Saudi Arabia

K. Pringle discussed the Kingdom of Saudi Arabia's initiatives under Vision 2030, aimed at empowering women in the energy sector. With strong leadership support, the Kingdom has made notable progress, including targeted programmes like the MIT fellowship for women in energy and the establishment of the NGO "Women in Energy". The Kingdom has also focused on increasing female participation in traditionally male-dominated roles, such as gas station inspectors, where female representation was pushed to 50%. These efforts are part of a broader strategy to integrate women into the growing nuclear and renewable energy sectors.

D. Kgomo, Co-Chair, Nuclear Regulators for Gender Equity (NRGE)

D. Kgomo emphasized the importance of regulatory frameworks in ensuring gender equity in the nuclear sector. NRGE focuses on creating opportunities for women through training, mentorship, and policy advocacy. She highlighted the success of mentorship programmes that have connected young female professionals with experienced leaders in the field, helping to bridge the gap in gender representation at higher levels of nuclear regulation.

A. Des Cloizeaux, Director of Division of Nuclear Power, IAEA Nuclear Energy Department

A. Des Cloizeaux discussed the role of the IAEA in promoting gender diversity within the nuclear industry. The IAEA is committed to promoting gender equality both internally and externally, with a goal of achieving gender parity in its secretariat by 2025. Since January 2022, the agency has made significant progress, with the proportion of women in technical roles increasing. The IAEA supports gender equality through various initiatives, such as a mentoring programme that pairs women with mentors, providing training to combat bias and build self-confidence. Externally, the Agency encourages women in STEM through programmes like the Marie Skłodowska-Curie Fellowship, particularly supporting women from developing countries. Additionally, the IAEA is developing tools, including a manuscript on human resource strategies for gender equality, to help member states implement equitable policies and foster leadership, employee engagement, and career development for women.

C. Thomas, Founder and CEO, Thomas Thor, UK

C. Thomas emphasized the evolving needs and dynamics in the nuclear sector's workforce, particularly in recruiting and retaining talent. He highlighted the growing importance of purpose, diversity, inclusion, and flexibility in attracting younger generations. C. Thomas underscored the shift from managing based on presence to focusing on outcomes, especially in the context of flexibility and remote work. He also pointed out the challenges of managing multi-generational teams and the critical traits necessary for effective leadership, such as communication, empathy, and adaptability. He advocated for empowering managers with the right resources and recognition to foster gender balance and diversity, stressing that successful organizations are those where leaders actively drive change. Finally, he redefined the role of allies in promoting gender balance, suggesting that allies need to be actively involved in enabling progress and supporting underrepresented groups.

The session also provided the opportunity to present a video on the STREAM-clusive project from Malaysia, – the IAEA proSTEM Challenge 2024 winner. The video was presented by K. V. Ket and accompanied by his comments on the project, that aimed at increasing female participation in STEM fields. The project focuses on creating inclusive educational content and providing mentorship to girls, children of colour and students with special needs, that are interested in pursuing careers in science, technology, engineering, and mathematics.

The Q&A part of the session focused on challenges and strategies related to advancing gender equality and women's empowerment in various sectors, particularly in the nuclear and energy industry. It underscored the ongoing challenges and diverse strategies required to advance gender equality in different cultural and organizational contexts, highlighting the importance of leadership support, cultural sensitivity, and continuous dialogue.

1. Backlash Against Gender Equality Initiatives

E. Pule shared that Eskom's women advancement programme faced backlash due to its aggressive approach. The organization addressed unconscious bias by focusing on leadership development for women and providing them with global exposure. Despite progress, unconscious biases and subtle resistance persist, highlighting the need for ongoing efforts to challenge patriarchal perceptions.

2. Expanding Gender Initiatives in the Middle East

K. Pringle discussed Saudi Arabia's efforts to ensure gender equality and sustain progress, particularly in the energy sector. She emphasized the importance of slow and sustainable steps, regional partnerships, and addressing cultural challenges to women's participation in the workforce. The Kingdom is making strides with new educational programmes for women in energy-related fields and engaging with international organizations to extend these efforts regionally. Some success examples in the region were also mentioned.

3. Supporting Work-Life Balance and Attracting Women to Remote Areas

K. Pringle stressed the importance of listening to women to understand their needs, as cultural differences shape their experiences. She noted that labour laws in the Middle East are often more progressive in supporting women's needs than in other regions.

C. Thomas highlighted the importance of visibility for role models, proactive outreach, and flexible work arrangements to attract more women to the sector, particularly in remote areas. He believes achieving a critical mass of gender balance will eventually make such discussions obsolete.

4. Managing Equity, Quality, and Empowerment

The final question addressed how to manage equity, quality, and empowerment in supporting women's careers. The moderator emphasized the need for dialogue at all levels, the critical role of male allies, and the importance of adapting strategies to different cultural contexts.

Closing statement was made by W. Huang, Director of Nuclear Energy Planning, Information and Knowledge Management, IAEA. He highlighted the significance of gender topics discussed during the conference, noting the progress made but also emphasizing the persistent gender gaps in the global professional world, particularly in the nuclear sector. He acknowledged the efforts of the IAEA, member states, international partners, and academia in addressing these challenges. Despite improvements, especially within the IAEA, W. Huang stressed the need for continued efforts to achieve gender parity. He concluded by encouraging boldness in young women and expressed gratitude for the successful discussions, wishing everyone a productive conference.

5. PEOPLE: DEVELOP, EMPOWER, LEAD

5.1. KEYNOTE SESSION

J. Isotalo
Senior Vice President, Teollisuuden Voima Oyj (TVO), Finland

J. Isotalo shared insights into Finland's nuclear energy journey, particularly the development of the Nuclear Professional Programme. She emphasized the long and sometimes difficult process but highlighted the value of learning through challenges. Finland, despite its harsh climate and small population, operates five nuclear reactors that contribute to 40% of its electricity, with 94% of the country's electricity being emission-free in 2023. The success of Finland's nuclear sector, particularly with the Olkiluoto 3 reactor, has allowed the country to transition from the largest electricity importer in the EU to self-sufficiency in clean energy production.

Olkiluoto is unique in the nuclear world as it encompasses all necessary nuclear waste management facilities on one island. This includes an interim storage facility for spent fuel, a repository for low and intermediate waste, and a geological disposal facility for spent nuclear fuel – the first of its kind in the world. This comprehensive approach to nuclear waste management has positioned Finland as a leader in the field. The success of Finland's nuclear energy industry is also bolstered by strong public support, with only 6% of Finns opposing nuclear energy, indicating a high level of acceptance for the sector.

J. Isotalo also discussed how the economic environment between 2010 and 2015 impacted TVO. The electricity price drop and challenges in the Olkiluoto 3 project forced the company to make major organizational changes, leading to staff reductions and a decrease in internal safety culture indicators. This shift alarmed both TVO and its international nuclear partners, prompting the company to seek external support from the International Atomic Energy Agency (IAEA) and the World Association of Nuclear Operators (WANO). Together, they identified leadership as a critical area for improvement and implemented various workshops and training programmes to address the issue.

One major realization for TVO was understanding how the national culture influenced leadership development. Initially, management expectations were unclear and not integrated into everyday practice, despite being visibly present in documents and posters. By addressing this, TVO was able to create a more engaged and accountable organizational culture, centred on nuclear professionalism. The introduction of five key characteristics for nuclear professionals and leaders helped to redefine the company's values and practices, leading to a significant cultural shift.

Since then, TVO has seen remarkable improvements. The company now has the highest-ever levels of employee satisfaction and safety culture indicators. TVO's strong recovery is evidenced by 200 people of increase in staff since 2015 and positive evaluations from IAEA and WANO. Additionally, the company has established the first green bonds in the EU's nuclear sector, raising substantial funding to support its clean energy goals. Olkiluoto' s operations are stable, and TVO is preparing for lifetime extensions and power increases for its older units.

In closing, J. Isotalo encouraged her international colleagues to actively engage with organizations like IAEA and WANO, sharing both successes and challenges openly. She emphasized that collaboration and learning from both positive and difficult experiences are key to the continued growth and safety of the nuclear industry.

T. Ivanova
Head of the Division of Nuclear Science and Education
Nuclear Energy Agency

T. Ivanova opened her presentation by introducing the NEA, which supports international cooperation among 34 advanced nuclear technology countries. The NEA's mission is to assist its members in developing the scientific, technical, and legal foundation for the safe and economical use of nuclear energy. She highlighted the NEA's 2022 report on nuclear energy's role in meeting climate targets, emphasizing the importance of long-term reactor operations, large reactors, and small modular reactors (SMRs) in the global transition to net-zero emissions. She stressed that training and education have to evolve to meet the demands of the rapidly expanding nuclear sector.

T. Ivanova underlined the importance of international cooperation in education and training within the nuclear sector. She noted that many nuclear suppliers are multinational, making global collaboration essential to developing a skilled, connected workforce. Workshops organized by the NEA with stakeholders from government, industry, and academia highlighted the need to enhance nuclear education. Increased interest in nuclear programmes among younger generations is promising, but there is still a need to create awareness of nuclear careers and secure sustainable funding. Early outreach to students, even before university, is crucial for building a future workforce.

T. Ivanova presented the results of NEA discussions with academia and industry. Universities need adequate faculty, modern infrastructure, and international cooperation to meet growing educational demands. Industry, on the other hand, needs professionals with both technical and soft skills who can communicate effectively with diverse stakeholders. The NEA's Nuclear Education, Skills, and Technology (NEST) framework and international collaboration have proven essential in providing hands-on experience to students. These initiatives help bridge the knowledge gap and ensure that research infrastructure remains available for future education and training.

T. Ivanova then addressed gender balance in the nuclear sector, highlighting the findings of the NEA's 2023 report. Women remain underrepresented, especially in STEM and leadership roles, with only 20% of the STEM workforce and 18% of senior leadership positions held by women. She stressed the need for visibility, lifestyle improvements, and a more supportive workplace culture for women in nuclear. She emphasized that increasing diversity is not just about fairness but also about driving innovation and productivity in the sector.

To improve gender balance, Ivanova advocated for coordinated policies that attract women to nuclear careers, retain them through supportive work environments, and promote them into leadership roles. She pointed to the NEA's long history in nuclear education, from early initiatives in the 1950s to the recent launch of the 2035 project, which aims to attract and train the next generation of nuclear experts. These initiatives are critical to addressing workforce challenges and achieving diversity in the nuclear sector.

She stressed that nuclear energy is increasingly recognized as a key player in achieving global net-zero emissions goals. With new investments and technologies emerging, now is the time to ensure that the nuclear sector attracts a diverse and highly skilled workforce to power future growth. She ended her speech by expressing optimism for the future of nuclear energy and its role in fostering creativity and innovation.

C. Yang
Deputy Chief Economic Officer and Chief Human Resource Officer
China National Nuclear Corporation, People's Republic of China

C. Yang presented CNNC's two innovative practices in human resource management (HRM). The first practice he discussed was the development of CNNC's Human Resource Management System Standards, which aim to create unified HRM evaluation methods and address gaps in macro-level standards. The speaker explained that many organizations rely on business metrics that lack connectivity and adaptability to external and internal challenges. To solve this, CNNC developed standards based on the total quality measurement system. These standards include 10 chapters and 30 sections covering areas such as leadership, planning, evaluation, and improvement.

Three key features of the standards include:

- Element-based approach: Identifying key elements that impact HRM quality and effectiveness and defining basic requirements across organizations;
- Systematization: Emphasizing that HRM is a cohesive system, not just a collection of separate modules;
- Continuous improvement: Using the PDCA (Plan-Do-Check-Act) cycle to ensure ongoing refinement of HRM processes.

C. Yang further elaborated on how CNNC applies these standards by adopting a process-based approach. HRM is treated as a series of interconnected processes, where the output of one process becomes the input for the next. He emphasized that all HRM activities, from planning to evaluation, follow the PDCA cycle, ensuring continuous improvement and systematization.

He also discussed the development of evaluation criteria for HRM, which include 12 primary indicators, 29 secondary indicators, and over 160 quantitative inspection clauses. Trained auditors assess HRM systems based on these criteria. The evaluations have been implemented across 130 CNNC units, and 52 of them have undergone systematic evaluations, leading to clear insights into areas for improvement.

The second innovative practice discussed was 'the Quantitative Appraisal Method for Management Department Employees'. This method was developed to accurately measure the performance contributions of employees in management departments. The appraisal is based on the concept of standard working hours and consists of the following steps:

- Systematic breakdown of work: Tasks are divided into modules, first-level tasks, second-level tasks, and basic units. CNNC's HRM system comprises 12 work modules, broken down into over 2,100 basic units;
- Determination of standard working hours: The time required to complete tasks is determined through observation, expert input, and employee discussions. These times serve as benchmarks for evaluating employee performance;
- Comprehensive work evaluation: Employees are assessed based on task performance, quality, efficiency, and collaboration. Their performance is quantified using a Q score and multiplied by their total working hours;
- Multidimensional application of results: Working hours data are published regularly to encourage a competitive mindset among employees. The method supports job assignments, staffing, and schedule optimization.

This method has enhanced organizational transparency by making employee performance measurable and comparable. Yang highlighted that the method has been recognized by the National Enterprise

Management Modernization Innovation Achievement and has been adopted by 100 organizations across multiple sectors.

In closing, C. Yang emphasized CNNC's commitment to sharing best practices and invited further collaboration to improve HRM capabilities. He concluded by thanking the audience and expressing CNNC's openness to exchange ideas with international colleagues.

5.2. PLENARY SESSION

M. Uesaka
Chairperson, Japan Atomic Energy Commission, Japan

M. Uesaka delivered a presentation on the topic of nuclear knowledge management, focusing on the development of professional engineers in the nuclear sector. He highlighted the challenges Japan faces due to a significant temporal gap in building and operating nuclear power plants (NPPs). For over 20 years, Japan did not engage in new nuclear construction, and after the Fukushima accident, half of the country's reactors remained inactive for a decade. This has led to a shortage of experienced middle-aged engineers with practical knowledge in nuclear plant operations. As a result, effective knowledge management is crucial, particularly in transferring expertise from senior to younger engineers, while also addressing gaps in the latter's technical education, as universities increasingly emphasize software and AI over traditional engineering.

To combat these challenges, the speaker discussed the role of Tokyo's new nuclear professional school, which prepares students to earn national licenses as chief nuclear operators and chief nuclear field engineers. The school has an enrolment capacity of just 15 students and collaborates extensively with the Japan Atomic Energy Agency (JAEA) and the nuclear industry. The students, mostly seasoned professionals with several years of industry experience, undergo rigorous education with a focus on deep technical knowledge and practical skills. M. Uesaka emphasized the serious, dedicated atmosphere at the school, where students engage in intensive study, asking questions and actively participating in discussions.

In addition to traditional classroom instruction, the University of Tokyo has developed comprehensive e-learning materials, including more than 10,000 slides that have been incorporated into the IAEA's cyber platform for nuclear education. The JAEC Chair provided a detailed overview of the partnership between the University of Tokyo and the IAEA, which began in 2010 and culminated in the university becoming an INMA-certified institution in 2020. This certification process took a decade, and M. Uesaka noted that the school has graduated a significant portion of Japan's licensed nuclear engineers over the past 20 years.

M. Uesaka also discussed the broader role of professional engineers in the nuclear field, drawing comparisons with other countries like the United States, the UK, and France, where professional licenses for engineers play a crucial role in ensuring the safety and quality of nuclear projects. He advocated for Japan to strengthen its own certification system for nuclear and radiation engineers to enhance global competitiveness. The key areas of competence for Japanese nuclear professional engineers include academic knowledge, problem-solving, leadership, and ethics.

Finally, M. Uesaka emphasized the need for international collaboration and standardization in developing nuclear professionals. He expressed his hope that the IAEA would continue to support initiatives like INMA and promote the recognition of professional engineers in the nuclear field on a global scale. He concluded by envisioning a future where the title of "nuclear professional engineer" would carry the same weight and prestige as PhDs, medical doctors, and MBAs.

J. Uzhakina
Director General, Rosatom Corporate Academy, Russian Federation

The head of Rosatom Corporate Academy provided an overview of the Academy's educational programmes and the importance of corporate universities in addressing key challenges in the nuclear industry. She highlighted three major challenges: the skill gap, lack of employee motivation, and the rapidly changing business and technology landscape. To bridge the skill gap, Rosatom has built an educational ecosystem that supports smooth talent transitions from schools to the nuclear industry. J. Uzhakina also emphasized the importance of employee engagement, particularly among Generation Z, by fostering open communication and shared values within the organization. Finally, she pointed to the need for continuous adjustment of training programmes to keep pace with technological and business changes, particularly in areas like digital skills and complex project management.

The Rosatom Corporate Academy, since its inception, has been instrumental in developing programmes to address these challenges. J. Uzhakina described the Academy as a key operator in Rosatom's strategic human resources projects, offering over 300 training programmes led by 600 trainers. These programmes reach a wide audience, including employees, students, and residents of Rosatom-operated cities. The Academy tailors its learning ecosystem to different target groups, such as young people, future leaders, technology experts, and engineers, ensuring the professional development and continuity of talent pipelines in the nuclear industry.

One of the Academy's main focuses is youth engagement and development, with approximately 120,000 employees under the age of 35. The Academy plays a pivotal role in shaping the career paths of young professionals entering the nuclear sector. To foster this engagement, Rosatom has established four youth communities: the Rosatom Junior Council, Student Council, Youth Council, and the Impact Team. These communities provide platforms for young people to engage in decision-making, promote STEM disciplines, and participate in nuclear outreach programmes on an international level.

In addition to youth programmes, Rosatom's leadership development programmes aim to nurture the next generation of industry leaders. J. Uzhakina outlined two key initiatives: a leadership programme, which prepares new leaders for strategic challenges, and a management school focused on practical skills like performance and project management. Over 40,000 people in leadership positions participate in these programmes, ensuring that Rosatom's management culture evolves alongside industry demands.

The Academy also focuses on reskilling and upskilling through its specialized technology schools, which train employees in fields such as IT development, construction project design, and additive technologies. These schools will involve up to 150,000 employees by 2030, with a certification system in place to ensure the development of critical skills. Another initiative, AtomSkills, focuses on training engineers and blue-collar workers through competitions and practical skill assessments. This flagship programme not only improves technical skills but also promotes vocational training as a desirable career path for younger generations.

In conclusion, J. Uzhakina stressed the importance of fostering essential skills such as critical thinking, digital literacy, and complex project management in all employees. She called for continuous dialogue and cooperation to develop effective training and development tools, ensuring that Rosatom remains at the forefront of industry innovation.

J. Dies
Commissioner, Nuclear Safety Council, Spain
Chairman of Spanish Nuclear Energy Technological Platform R&D (CEIDEN)
Professor Chair in Nuclear Engineering

The presentation focused on the development of nuclear skills and human resource management within nuclear regulatory bodies. Drawing from his 38 years of experience, J. Dies emphasized that the safe and sustainable use of nuclear technology relies heavily on a highly educated workforce with strong knowledge in nuclear engineering, radiation protection, and nuclear safety. He also highlighted the global trend of extending the operation of nuclear power plants to 80 years and the ambitious plans of 33 countries that have announced that they will build 495 new nuclear power plants (WNA), further emphasizing the need for specialized nuclear professionals.

One of the key points was the importance of recruiting skilled individuals into the nuclear sector, and in regulatory bodies. The speaker discussed the need for strong connections between universities offering nuclear engineering programmes and the nuclear industry. He stressed that attracting students into these programmes and ensuring that they remain in the nuclear field, rather than pursuing careers in unrelated sectors, is crucial for maintaining a competent workforce. Establishing university programmes that produce nuclear engineers who then join the regulatory body and the nuclear sector is an essential part of this effort.

J. Dies provided examples of successful programmes in Spain that have been used to build a strong educational pipeline for the nuclear industry. He mentioned the European Master of Science in Nuclear Engineering certification and the importance of collaboration between universities, regulatory bodies, and the nuclear industry. In Spain, the Nuclear Safety Council has sponsored chairs in nuclear safety and radiation protection at universities, supporting education and ensuring the quality of nuclear engineering programmes. Additionally, partnerships with nuclear power plants and companies have allowed students to gain practical experience through internships and lectures delivered by industry professionals.

He also highlighted good practices for recruitment and development of talent, such as the establishment of nuclear employment portals, which connect nuclear graduates with job opportunities in the nuclear sector. These platforms ensure that students are informed about job openings as soon as they arise. Furthermore, J. Dies emphasized the importance of consistent hiring to maintain a stable influx of talent in the sector, rather than fluctuating numbers that may leave gaps in staffing.

Another critical area discussed was the need for continuous support to universities with well-established nuclear education programmes, both nationally and internationally. He suggested that collaborations between universities, nuclear regulatory bodies, and international organizations like the International Atomic Energy Agency (IAEA), the Nuclear Educations Networks, are key to maintaining high standards in nuclear education. The speaker advocated for more scholarships and funding to support students in nuclear programmes, especially in Europe, where recruiting nuclear engineers can be challenging.

In conclusion, the speaker presented data estimating the hiring needs for the Spanish nuclear sector over the next five years, which includes 1,400 graduates and 700 vocational training technicians. He emphasized the importance of strategic planning in human resource development to meet the future needs of the nuclear industry, ensuring that both highly educated engineers and skilled technicians are available to support the growth and safe operation of nuclear facilities.

5.3.   SUMMARIES OF THE PARALLEL SESSIONS

5.3.2. Session 5.1. Workforce Development for New Nuclear Power Projects

Session Chair: J. Gadano, Argentina

IAEA Session Rapporteur: T. Reysset

The session explored critical aspects of building and enhancing workforce capabilities to support emerging nuclear project initiatives. Topics included Canada's efforts to expand its workforce for new SMR deployments, capacity building in newcomer countries like Egypt's El Dabaa NPP, Croatia's approach to effective NKM, methodologies such as SAT for skill unification in Westinghouse, and Sweden's SKB approach to competence development and planning. Collaboration with institutions and the IAEA is pivotal in shaping effective strategies for workforce development in these endeavours.

Paper ID 450: Developing a National Nuclear Workforce Development Framework in Canada

Speaker: D. Tucker, Canada

The presentation describes the new workforce challenges faced by Canada due to accelerated nuclear projects and GHG-free energy commitments, including SMR development and expansion into new provinces. To meet these demands, Canadian organizations are enhancing workforce development through initiatives like expanded nuclear education, industry-based training programmes, mentorship from experienced professionals, and strengthened management training.

Paper ID 263: Personnel Training for First NPP in Country

Speaker: D. Podoliakin, Russian Federation

The presentation describes Rosatom approach to capacity building in a newcomer country, Egypt. Key projects include training and licensing personnel for Egypt's first NPP, El Dabaa. Challenges include advanced training needs, limited experienced personnel, and ensuring qualifications. To address these, long-term, staged training programmes with supervised competence development are essential, taking up to 4 years for some positions. Effective training requires technical equipment to meet objectives. The presentation outlines an approach to meet SAT requirements, ensuring timely personnel training for the first NPP.

Paper ID 113: Opportunities and Challenges in the Development of National Nuclear Knowledge Management Programme

Speaker: S. Pleslic, Croatia

The paper presents Croatian approach to effective Nuclear Knowledge Management. Key focuses are safety, security, and sustainability. Effective Nuclear Knowledge Management (NKM) requires addressing people, processes, technology, culture, and structure, with input from various institutions and the government. The goal is to find optimal methods for managing and sharing knowledge. Organizational culture and structure are crucial for NKM implementation, which should be evaluated through a SWOT analysis. The paper also reviews nuclear capacity building activities.

Paper ID 338: Knowledge Management and the Global Competency Model at Westinghouse

Speaker: F. Ruiz Martinez, USA

The presentation introduces Westinghouse knowledge management methodology to support competencies management. The SAT methodology enhances the identification of these competencies. WEC's team specializes on SAT, optimizing management processes and preserving knowledge for safe facility operations. They developed the Global Competency Model (GCM) tool for qualification management, skill analytics, employee training, and knowledge requirements. The GCM fosters technical development, delivers vital information, and builds partnerships.

Paper ID 429: From strategic to operational competence planning and processes

Speaker: J. Gustafsson, Sweden

The paper presents SKB approach to competence development and planning. It presents how different areas of planning and processes within NKM and HR (such as critical competence, SAT and competence transfer) connect to attract, retain and enhance employees, in the short and long term. It also presents an example of how working with "70-20-10" model and "before, during and after" training, together with focus on competence, behaviour and effects helps an organization to reach a successful road to enhanced competence over time.

5.3.3. Session 5.2. NKM-HRD in Fusion Organization

Session Chair: R. Kamendje, EUROfusion

IAEA Session Rapporteur: J. Roberts

The session focused on the importance of knowledge management (KM) in advancing nuclear fusion, particularly magnetic confinement fusion, and strategies for capturing and transferring tacit knowledge. Presentations covered the efforts of various organizations. EUROfusion highlighted its integration of EU fusion research, low gender diversity (21%), and upcoming initiatives, including the launch of the Fusion Education and Learning Hub (FuEL) and a new KM strategy. The UKAEA discussed its formalization of a Knowledge and Information Management (KIM) strategy to address an aging workforce and siloed communication, drawing on best practices from organizations like NASA. Fusion for Energy (F4E) detailed its KM strategies to manage knowledge flow and address challenges such as an aging workforce and secure knowledge transfer within the ITER project.

Presentation: The Current Nuclear Fusion Landscape

Speaker: R. Kamendje, EUROfusion

The presentation summarizes the current nuclear fusion landscape focusing on magnetic confinement fusion. KM has been identified as a key in the push for commercialization.

Paper ID 112: The EUROfusion Knowledge Management and Human Resource Development Strategy

Speaker: E. Belonohy, EUROfusion

The paper presents the EUROfusion Knowledge Management Strategy based on a four-pillar model, namely converting tacit to explicit knowledge (knowledge capture), tacit to tacit knowledge transfer between experts (community building), building tacit knowledge (training and education) and ensuring knowledge is accessible, used and embedded into everyday processes (tools and accessibility). E. Belonohy briefly explained the EUROfusion consortium, it integrates the fusion research in the EU funded by the EU Commission. EUROfusion publishes their Human Resource Survey every 10 years, the most recent survey was completed in 2023. Gender diversity is only 21% which is similar to CERN, ESO and ESA. This year the number of female applicants has been doubled on some programmes. There are 100 universities affiliated to EUROfusion with 829 PhD and 291 MSc students, some financially supported by EUROfusion. Career path events are organised by EUROfusion. The Fusion Education and Learning Hub (FuEL) will launch in September 2024 with 3 live courses and 6 recorded courses. A KM strategy will initiate some new initiatives in 2024-25. Generally, more technicians, engineers and operators are needed. ITER has recently hosted an event to discuss knowledge transfer to the commercial fusion sector.

Presentation: UKAEA Knowledge and Information Management Programme

Speaker: K. Carr, United Kingdom

UKAEA's mission is to lead the delivery of sustainable fusion energy and maximize the scientific and economic benefit. UKAEA have a 60-year journey to capture and preserve information from research, design, build, test and now decommissioning. To this effect, a knowledge and information management (KIM) programme has been set up. The paper presents the strategy at the core of this programme. K. Carr discussed the knowledge and information management challenges at UKAEA. An aging workforce with a 60-year history prompted the formalisation of a Knowledge and Information Management (KIM) strategy in 2023. It is based on previous good practices from organisations such as NASA. The IAEA NKM publications were reviewed but UKAEA felt they were too advanced for their initial level of KM. The UKAEA was working in silos with little communication across the site at Culham. Following a KIM audit, a KM Process will be implemented.

Presentation: Strategies for bridging the generational gap in the ITER Project. Key insights from F4E

Speaker: P. Mahedero, Fusion for Energy

As the organization responsible for the delivery of Europe's contribution to the ITER project, Fusion for Energy is required by the French regulator to implement knowledge management throughout the organization. This paper focusses on the strategies for bridging the generational gap in the ITER Project. P. Mahedero discussed the KM needs, challenges and strategies of Fusion for Energy which is the co-ordinated EU support for the ITER project. The challenges include facilitating secure knowledge flow between ITER departments and partners. Many of the employees are reaching retirement age. F4E have developed a range of KM strategies.

5.3.4. Session 5.3. Establishing Training Programmes

Session Chair: G. Bikkulova, Russian Federation

IAEA Session Rapporteur: R. Kvetonova

The crosscutting topic of the session was how to deal with the challenges related to an ageing workforce. It has also emphasized the role of Nuclear Knowledge Management in supporting most business functions, including HRD, training and risk management. It has been discussed in several presentations that education and training are key tools for preserving and sustaining knowledge. Rosatom introduced its activities in hosting of Joint Russia-IAEA Nuclear Energy Management and Nuclear Knowledge Management Schools to support IAEA Member States in maintaining and preserving nuclear knowledge and various training methods, including face-to-face, online, and hybrid training, as well as practical classes with 3D modelling, videos, and virtual classrooms.

Paper ID 153: Implementation of Nuclear Knowledge Management Programme on a New Nuclear Facility

Speaker: H. Elsayed, Egypt

The paper underscores the significance of Nuclear Knowledge Management (NKM) in the nuclear energy industry. It highlights NKM's role in supporting various business functions, including human resource management, training, planning, operations, and risk management. The paper addresses challenges related to an ageing workforce and emphasizes the need for formal approaches to manage nuclear knowledge and skills. Strategies for capturing, retaining, and transferring nuclear expertise are discussed. Finally, it touches upon the implementation of a knowledge management programme in a new nuclear facility.

Paper ID 249: Nuclear Knowledge Management: The role of Centre for Energy Research and Training Zaria, in Education and Training of a Nuclear Workforce

Speaker: A. Sa'id, Nigeria

The paper discusses the importance of education and training in the sustainability of the nuclear industry. It highlights the role of the Centre for Energy Research and Training in transferring and sharing nuclear education in Nigeria. The paper emphasizes that education and training are key tools for preserving and sustaining knowledge. It also discusses the role of Nuclear Knowledge Management (NKM) and Human Resource Development (HRD) in building and sustaining capacity in nuclear organizations to support national nuclear programmes. The abstract further discusses new NKM and HRD approaches to ensure a skilled workforce for the future and to support the sustainable development of advanced technologies.

Paper ID 258: Rosatom Technical Academy Experience of Organizing Training Events under the IAEA's Technical Cooperation Programme

Speaker: E. Bologov, Russian Federation

Since 2017, Rosatom Technical Academy has been conducting training events under the IAEA Technical Cooperation Programme to develop nuclear energy infrastructure. These events have attracted participants from 59 countries, including both newcomers and countries with developed nuclear power programmes. Rosatom Tech annually hosts Joint Russia-IAEA Nuclear Energy Management and Nuclear Knowledge Management Schools to support IAEA Member States in maintaining and preserving nuclear knowledge. The training events involve highly qualified speakers and subject matter experts from leading Russian nuclear organizations. Participants gain theoretical knowledge and practical skills through classroom lectures, hands-on sessions, and site visits to nuclear facilities. Rosatom Tech's participation in the IAEA's Technical Cooperation Programme contributes to promoting the peaceful use of nuclear technology and strengthening nuclear safety and security worldwide.

Paper ID 316: The Importance of Education and Training in Operation of a Nuclear Power Plant Krško

Speaker: T. Gelo, Croatia

The paper discusses the crucial role of continuous training and development in ensuring the safe and stable operation of the Krško Nuclear Power Plant (NPP) in Slovenia. The NPP, a Westinghouse pressurized water reactor with a capacity of 700 MW, provides significant power to both Slovenia and Croatia. The paper emphasizes the importance of long-term human resource planning, timely staff recruitment, and systematic development for all employees in maintaining regulatory requirements and a culture of safety. It also highlights the low percentage of female staff and the principle of parity for personnel from Slovenia and Croatia in leading positions. The abstract concludes by discussing the reinforcement of knowledge and transfer of skills from experienced staff to younger generations through on-the-job training programmes and mentorship.

Paper ID 280: Practice-oriented Training of Foreign Specialists in Nuclear Security at the Rosatom Technical Academy

Speaker: E. Bologov, Russian Federation.

The Rosatom Technical Academy's Global Nuclear Safety and Security Institute (GNSSI) is a leading institution for training specialists in nuclear and radiation hazardous facilities. It has been an official IAEA training centre since 2019, focusing on practical skills and abilities. The Academy regularly conducts courses in nuclear security, information security, and accounting and control of nuclear materials. These courses combine advanced theoretical materials and applied modules, using modern training infrastructure. The GNSSI uses various training methods, including face-to-face, online, and hybrid training, as well as practical classes with 3D modelling, videos, and virtual classrooms. The report emphasizes the combination of theory and practice during training.

Paper ID 229: Talent Cultivation for Global Nuclear Energy Development: Contributions and Practices from China

Speaker: Y. Wang, People's Republic of China

The paper underscores the significance of Nuclear Knowledge Management (NKM) in the nuclear energy industry. It highlights NKM's role in supporting various business functions, including human resource management, training, planning, operations, and risk management. The paper addresses challenges related to an ageing workforce and emphasizes the need for formal approaches to manage nuclear knowledge and skills. Strategies for capturing, retaining, and transferring nuclear expertise are discussed. Finally, it touches upon the implementation of a knowledge management programme in a new nuclear facility.

5.3.5. Session 6.1. People, Knowledge and Operators

Session Chair: K. Mrabit, Morocco

IAEA Session Rapporteur: R. Kvetonova

The session's contributions were focused especially on different KM techniques and methods, i.e. managing critical knowledge and the assessment's impact. The presentations emphasized the importance of continuous evaluation and improvement of the training system. Six-dimensional strategy for human resource management, focusing on systematization, standardization, institutionalization, process-demonstration, digitalization, and detail-refinement has been also introduced. The importance of the Systematic Approach to Training (SAT) methodology in developing a reliable training management system has also been highlighted. The session discussed the importance of intercultural communication in Nuclear Knowledge Management (NKM).

Paper ID 298: Approach to identify and manage critical knowledge at Kozloduy NPP

Speaker: I. Cvetkov, Bulgaria

The Knowledge Management (KM) process at Kozloduy NPP is primarily focused on employees who possess undocumented knowledge that is critical to the company. The process involves defining key knowledge domains, prioritizing job positions based on their impact on safety, reliability, economy, and reputation, and focusing on employees with a high probability of knowledge loss. The KM process also includes identifying critical knowledge with a high risk of loss and planning preventive measures for knowledge retention. The company uses various techniques and tools to implement these preventive measures. An impact assessment is performed for each at-risk knowledge area to calculate the Criticality ratio for each identified knowledge/skill at risk. This assessment is necessary as it defines the risk of individual knowledge loss, considering the extent to which this knowledge is critical for the organization. The paper concludes by emphasizing the importance of continuous evaluation and improvement of the training system.

Paper ID 455: Promoting High Speed Development of Nuclear Power through Standardized Human Resources Management System

Speaker: G. Chen, People's Republic of China

Sanmen Nuclear Power Plant developed a six-dimensional strategy for human resource management, focusing on systematization, standardization, institutionalization, process-demonstration, digitalization, and detail-refinement. The company launched programmes to attract talented people, such as "practice bases for college students", "the elite programme", and "the talent enclaves". They introduced human resources from multiple channels, including "joint training", "school-enterprise cooperation", and "overseas talents recruitment". Platforms were built to tap the "hard skills" of talents, including "master-apprentice", "skill competition", and "talent declaration". The company also worked out "magnetic" policies to retain talents, including "supplementary medical care", "salary incentives", and "talent priority channels". The paper emphasizes the importance of building and sustaining capacity in nuclear organizations to support national nuclear programmes, nuclear education, and new approaches to ensure a skilled workforce for the future. It also discusses the role of HRD in supporting the sustainable development of advanced technologies.

Paper ID 260: Multi-step NKM and HRD Process in the Paks 2 Project

Speaker: Z. Bács, Hungary

The paper discusses the challenges of Nuclear Knowledge Management and HR Development in the nuclear industry, particularly for nuclear power plants under construction. It presents the case of the Paks 2 NPP project in Hungary, where the training management system had to be developed in the absence of final documents and personal conditions. The paper highlights the importance of the Systematic Approach to Training (SAT) methodology in developing a reliable training management system. It also discusses the specific aspects that need to be considered when the nuclear power plant is in the design and construction life cycle phase. These include harmonizing the schedules of classroom trainings and commissioning works, changes in Systems, Structures, and Components (SSC) during commissioning, and the availability of operational manuals. The paper concludes by discussing the cyclical scheme of the SAT methodology used in the Paks 2 NPP project, which ensures the training and readiness of personnel for the Commissioning and Start Up of the new units. The paper emphasizes the importance of continuous evaluation and improvement of the training system.

Paper ID 257: Intercultural Communications as an Element of NKM: Language Barriers and Ways to Overcome Them

Speaker: V. Mitinskaya, Russian Federation

The paper discusses the importance of intercultural communication in Nuclear Knowledge Management (NKM). It highlights the challenges faced due to language barriers and cultural differences during project realization in partner countries. The paper presents ROSATOM's experience in mitigating miscommunication risks at the early stages of project realization. It emphasizes the need for training and education in the nuclear industry and the role of language proficiency in expanding opportunities. The paper also discusses the psychological and linguistic aspects of language barriers and the need to overcome these for effective communication. It outlines the tasks faced by specialists, such as developing language, speech, and communicative competencies among students, and implementing career guidance activities. The paper concludes by discussing two programmes developed by Rosatom and the State Institute of Russian Language named after A.S. Pushkin, aimed at training foreign students and schoolchildren in the Russian language for nuclear and related areas.

Paper ID 227: Hitachi-GE Nuclear Energy, Ltd.'s Nuclear Knowledge Management for Transferring Technical Knowledge to Next Generation

Speaker: M. Ono, Japan

Hitachi GE Vernova Nuclear Energy, Ltd. (Hitachi-GE) initiated a corporate-wide Knowledge Management (KM) programme in 2017 to address the potential loss of knowledge and skills from senior engineers retiring. The company identified approximately 16,000 knowledge domains and experts who possessed each technical knowledge. A systematic approach was used to transfer this critical knowledge from veterans to younger generations through various methods such as codification and organization of core documents, workshops, and capturing tacit knowledge via videos. A robust KM governance was established, including management reviews, setting and monitoring key metrics, and promotional activities like an annual "KM Award". The commitment and leadership of top management were key drivers of this journey. The company also developed a Knowledge Transfer System (KNOTS) to efficiently manage the knowledge transfer process. Hitachi-GE is now considering the use of advanced technologies like generative AI and the metaverse for future KM activities. They also plan to implement a mentoring programme focusing on the transfer of tacit knowledge to cultivate the ability of the younger generation to deal with complicated situations and unknown problems.

Paper ID 9: Knowledge Loss Risk Analysis: Recognizing the Criticality of People's Abilities and Skills

Speaker: P. Lopez, Argentina

The study focuses on assessing the risk of knowledge loss due to retirements within organizations, specifically in the nuclear field. It introduces a methodology for evaluating this risk, considering factors such as the difficulty of finding replacements and the criticality of the roles. The research was applied to the National Atomic Energy Commission (CNEA) of Argentina, revealing that 4% of the population faced very high risk and difficulty of replacement, primarily in nuclear-related departments. Based on the findings, a role-based procedure for risk analysis was developed, emphasizing strategic functions and specialization levels. Additionally, a step-by-step guide for knowledge retention was designed, including systematic staff exit analysis, replacement examination, and training plans.

5.3.6. Session 6.2. Safety and Security Culture Including Capacity Building Aspects for Security

Session Chair: H. Looney, IAEA

IAEA Session Rapporteur: A. Ganesan

The session featured six presentations that explored key aspects of leadership, education, and capacity building in nuclear safety and security. One presentation highlighted the development of experiential learning through an international school for safety leadership, while another focused on enhancing safety culture through leadership education programmes and professional training. A strategic approach to capacity building in nuclear safety and security was discussed, emphasizing the development of human resources and competencies. Knowledge retention and transfer within nuclear security were also covered, along with strategies for improving nuclear security education and training. The session concluded with insights into the use of knowledge management tools for addressing cybersecurity challenges in advanced nuclear reactor technologies. Overall, the session underscored the critical importance of capacity building to ensure safety and security in the nuclear sector.

Presentation: The International School on Nuclear and Radiological Leadership for Safety

Speaker: M. Moracho Ramirez, IAEA

The Agency and its Member States have recognized the importance of leadership for safety in its fundamental safety principles, and in particular Principle 3, which states that "Effective leadership and management for safety must be established and sustained in organizations concerned with, and facilities and activities that give rise to, radiation risks." The Agency has also published a safety requirements document, GSR Part 2 that sets out under Requirement 2 the need for "demonstration of leadership for safety by managers".

Since 2016, research and development for the concept of an IAEA International School for Nuclear and Radiological Leadership for Safety took place culminating in an IAEA International School of Nuclear and Radiological Leadership. The overarching objective of the International School for Nuclear and Radiological Leadership for Safety is for early to midcareer professionals to develop their safety leadership potential through a better understanding of what leadership for safety means in practice in nuclear and radiological working environments with their inherent complexities and often competing considerations.

The presentation highlighted the experience of developing an experiential based learning innovative methodology and conducting more than 22 Leadership schools all over the world. Also, opportunities for cooperating globally at academic level and with educational institutions to address the challenges of developing the necessary human resources worldwide are pointed out.

Paper ID 129: Leadership for Safety Education: A Key to Successful Integrated Management and Safety Culture Enhancement

Speaker: J. Repussard, France

This paper begins with the three safety requirements to push back on organizational and human limits and the following three interlinked challenges:

- Getting used to new types of requirements, addressing organisations and their managers: the temptation of a performative approach;
- Implementing qualitative requirements: the temptation of over-relying on artefacts (reporting systems);
- Adjusting regulatory oversight: the temptation of developing extensive rules & guidelines to facilitate inspection.

To address the above challenges the paper highlights the role of capacity building for the managerial workforce, especially at the lower echelons of organizations. It further elaborates the following approaches as a solution to the challenge.

- Professional Training initiatives: IAEA, WANO;
- ELSE/DMaLSE: two complementary Master Level University Diplomas.

Paper ID 225: Leadership and Culture for Safety and Security Basis of Regulatory Capacity Building: AMSSNuR Experience and Lessons Learned

Speaker: T. Marfak, Morocco

The presentation begins with the nuclear solutions for Morocco's socio-economic development, which are largely dominated by medical applications, in addition to ambitions for new nuclear energy programmes. To ensure a high level of safety and security of these nuclear applications, the government adopted, in 2014, the law 142-12 on nuclear and radiological safety and security and the creation of a regulatory body 'AMSSNuR' in charge of nuclear safety, security and safeguards. To accompany safe, secure, and sustainable nuclear applications, AMSSNuR has established an integrated strategic approach to develop and maintain the main capacity building pillars (Ref. 1) especially human resources, necessary knowledge and competencies, education and training programmes and international cooperation. According to IAEA requirements and orientations, AMSSNuR adopted several tools developed by the Agency such as SARCoN and the systematic approach for training (Ref. 2-3).

AMSSNuR conducted a study aiming to establish a national strategy for education and training in nuclear and radiological safety and security which permitted AMSSNuR to identify number of persons to be trained or retrained and to establish training programmes adapted to the identified needs.

In its approach, AMSSNuR considered regional cooperation with African countries including training programmes which could be developed within the framework of the FNRBA, AFRA, ANNuR, IAEA, the EU, and bilateral agreements.

Paper ID 172: Nuclear Materials Integration Lessons Learned in Nuclear Knowledge Management

Speaker: M. Gant, United States of America

The paper highlights the efforts of National Nuclear Security Administration's Nuclear Materials Integration Division (NMID) and their strategic approaches to retain and transfer knowledge throughout the NNSA's Nuclear Security enterprise. NMID has tested multiple methods to capture knowledge: a knowledge management plan, knowledge preservation catalogue, and various strategic communications documents including fact sheets for awareness, ad hoc analysis, and recommendation papers. The paper points out the following as the most effective qualities of the activity:

- In-person interaction with experts;
- Small group environments for training;
- Real-time correction/verification;
- Exposure to information at high capacity.

Paper ID 78: Nuclear Security Education and Training Capacity Development at the University of Port Harcourt: Outcomes and Prospects

Speaker: A. Kuye, Nigeria

The paper focuses on the strategies and methods that the Centre for Nuclear Energy Studies (CNES), University of Port Harcourt used to develop capacities and competences in nuclear security education and training. From its inception in 2010, CNES has actively engaged its senior academic staff to attend organized workshops on professional development courses in nuclear security on a train-the-trainer basis. To date CNES has: (1) upgraded the nuclear security content in its existing master's degree programme in nuclear engineering; (2) commenced a postgraduate certificate programme in nuclear security science (PGC-NSS) in 2017 and graduated twenty five students; (3) trained over 50 Nigerian and international officials in nuclear security; (4) received sponsorship and support from the US Department of State's Partnership for Nuclear Threat Reduction (PNTR); (5) set up an E-Classroom for distance learning in 2017; (6) secured an interim partnership with the Nigeria Atomic Energy Commission for funding of the PGC-NSS for three academic sessions; and (7) joined the International Nuclear Security Education Network (INSEN) in 2016 and received radiation measurement equipment as donation from the International Atomic Energy Agency. Finally, the envisaged areas for growth and improvement in nuclear security education at CNES are presented.

Paper ID 52: Strengthening the Knowledge Management Enabler in Nuclear Computer Security Planning Response Actions of Nearly Autonomous NPP through Hidden Markov Model

Speaker: D. H. Nugroho, Indonesia

Indonesia's electricity demand in 2060 will be supplied partly by nuclear power plants. Most of the power reactors offered to Indonesia are SMRs and microreactor type designs utilizing emerging, advanced new technology. Usually, many such advanced technologies implement online systems, increasing the vulnerability to attacks on security systems. Several microreactors are designed using nearly autonomous control systems in which operators unfamiliar with computer security may have difficulties in planning response actions in anticipation of a potential attack. To overcome this challenge, Knowledge Management (KM) enablers consisting of technology infrastructure and people competency can be utilized. The proposed KM enables creation of new knowledge for regulatory bodies in nuclear computer security planning response actions of nearly autonomous NPP. The new knowledge of operator response actions follows the third-level Sheridan method of autonomous control based on the Hidden Markov Model which responds at runtime. Sheridan's third level is considered to have optimal performance in this case. This model uses online parameters updating rules which make it better suited to adapt to changes in the cyber environment. This new knowledge is also important for regulating related technology by the regulatory body. The new knowledge was captured through assessment and knowledge sharing through cooperation with competent universities and national/international TSOs.

5.3.7. Session 6.3. Knowledge Transfer

Session Chair: O. Glöckler, IAEA

IAEA Session Rapporteur: M. Ovanes

The session highlighted the ongoing efforts by Member States to enhance nuclear knowledge and skill development, with a strong focus on mentoring, knowledge transfer, and overcoming challenges like resistance to change. Initiatives such as Ethiopia's emphasis on preserving radiation protection expertise, the UEA's Irshaad Youth Programme guiding young professionals, and EDF's knowledge transfer methods illustrate a diverse approach to human capital development. Indonesia's strategies to address senior resistance in knowledge transfer and Korea's mentoring experiences, shared through the IAEA, further underscore the importance of knowledge management. The session emphasized the critical role of fostering a learning culture, collaboration, and innovation in the nuclear sector to ensure continuity and future readiness.

Paper ID 178: Know-How Transfer Method

Speaker: O. Dandois Balthazard, France

The paper discusses the "Know-How Transfer Method," a systematic approach implemented by EDF to capture and transfer the invaluable vocational experience of employees involved in constructing EDF's nuclear reactor fleet. The method aims to ensure the continuity of critical knowledge and prevent activity disruption by implementing tri-party interviews, knowledge transfer plans for experts, training for facilitators, and an accelerated transfer in response to urgent departures to prioritize and transfer critical knowledge within a shortened timeframe of 3 to 6 months. This structured approach proved essential for preserving and leveraging the expertise of seasoned professionals, ensuring that both successes and failures are documented and leveraged for future growth, thus contributing to a robust learning organization and ongoing development and competence within EDF.

Paper ID 277: Implementation of Mentoring Programme to Facilitate Knowledge Management and Technical Transfer

Speaker: K. S. Kang, Republic of Korea

The presentation highlights some of the outcomes of a mentorship programme that the IAEA launched in 2009 under its Inter-regional Technical Cooperation project. Korea Hydro & Nuclear Power has been a significant contributor to this initiative by setting up mentorship courses to disseminate Korea's knowledge and experience. Over a span of nine years, KHNP has effectively implemented these mentorship programmes, involving 134 mentees from 32 Member States. The participants rated the course highly, with a satisfaction score of 4.7 (94%). The challenges faced during the mentoring and coaching process are discussed.

Paper ID 376: The Challenge of Knowledge Management for Preserving Radiation Protection Capabilities in Ethiopia Technology Authority

Speaker: E. T. Zege, Ethiopia

The paper highlights the importance of managing and preserving knowledge among radiological and nuclear professionals, and the need for robust KM practices within the Ethiopia's regulatory body. Through a survey rolled out to all professional employees, significant challenges were identified, such as gender disparity, an aging workforce, and varied levels of awareness about safety regulations. Despite a generally positive perception of the work environment and career advancement opportunities, these issues persist. Recommendations include establishing training and mentorship programmes, implementing recognition programmes, supporting mental health, and forming strategic partnerships with international organizations, to enhance knowledge acquisition and transfer and address employee retention and professional development within the regulatory body.

Paper ID 217: The Role of Knowledge Management in the Development and Deployment of Innovation in the Nuclear Sector

Speaker: C. Bright, United Kingdom

The paper explores the critical role of Knowledge Management (KM) in fostering innovation within the nuclear sector. It emphasizes that effective innovation requires the sharing of diverse perspectives, a culture of exploration and challenge, and a robust evidence base to support a precautionary approach. The authors propose an integrated framework combining KM processes and tools, the development of a collaborative learning culture, and systematic data collection to substantiate innovation. This framework is illustrated through two case studies: the development of the UK Office for Nuclear Regulation's innovation hub and the KM programme for the Moorside nuclear new build project. The paper concludes that KM is essential for reducing uncertainty and enhancing organizational capability, ultimately supporting safe, secure, and effective innovation in the nuclear industry.

Paper ID 427: Irshaad Youth Programme

Speaker: G. Balhamar, United Arab Emirates

The paper presents the Irshaad Youth Programme (IYP), a pioneering mentorship initiative by the Federal Authority for Nuclear Regulation (FANR) in the UAE. Led by the Education and Training Department in collaboration with the Youth Working Group, the programme aims to cultivate the next generation of leaders in the nuclear and radiation sectors through structured mentorship and continuous development. By pairing young employees aged 18 to 35 with senior management mentors, the IYP supports the development of various competencies – including legal basis, nuclear safety, radiation protection, excellence, leadership, communication, decision-making, and strategic thinking – to ensure a skilled cadre of nuclear professionals.

Presentation: Critical Knowledge Capture and Transfer

Speaker: M. Ovanes, IAEA

The presentation focuses on methodologies for capturing and transferring critical knowledge in nuclear organizations. It outlines the importance of identifying significant and critical knowledge, especially knowledge that, if lost, poses risks to organizational safety, performance, and commercial viability. The presentation emphasizes distinguishing between tacit and explicit knowledge, with a focus on turning tacit knowledge (gained through experience) into structured, actionable forms. Various techniques are discussed for capturing and transferring knowledge, including the importance of engaging experts and developing knowledge transfer plans. The presentation highlights the dynamic nature of critical knowledge and the need for continuous monitoring and mitigation plans to manage risks effectively.

5.4. SESSION ON LEADERSHIP

Session was moderated by P. Diéguez Porras, IAEA.

Keynote

A. Duncan, Department of Energy, Office of Legacy Management, USA

A. Duncan emphasized the critical role of leadership in building capacity within the nuclear industry, distinguishing it from traditional human resource management. She argued that leadership is an ongoing process of growth and learning, not limited to those with formal titles but accessible to anyone, regardless of background. A. Duncan highlighted that true leaders possess qualities like communication, integrity, and the ability to inspire, which are essential for navigating the challenges of the nuclear sector. She stressed that while technical skills are important, the industry also needs to prioritize developing leaders who can effectively communicate, engage stakeholders, and manage public perception. This preparation, she noted, is vital for ensuring that the next generation of nuclear professionals is equipped not just with technical expertise but also with the critical soft skills needed to lead.

A. Duncan also underscored the importance of diversity and inclusion in leadership, advocating for policies that support women and other underrepresented groups in balancing career and family life. She called for a leadership approach that welcomes people of all backgrounds, genders, and ages, viewing diversity as a strength that can enrich the industry. By creating an inclusive environment and providing necessary support, the nuclear sector can ensure that its workforce is ready to meet future challenges, particularly with the development of advanced nuclear technologies. A. Duncan concluded by expressing optimism about the future, urging the industry to embrace a broad and inclusive vision of leadership to drive progress and innovation.

Moderated Panel Discussion

Panel discussion was moderated by L. Lande, IAEA.

Panellists:

W. Ndubai, Director for Strategy & Planning, NuPEA, Kenya

G. Bikkulova, Deputy Director General, Rosatom Corporate Academy, Russian Federation

K. Pringle, Director of Human Capacity Building, K.A. CARE, Saudi Arabia

A. Al Mur, Human Resource Director, FANR, United Arab Emirates

P. Diéguez Porras, Head, Nuclear Knowledge Management Section, IAEA

Panel discussion focused on the challenges and opportunities in leadership within the nuclear sector, particularly in the context of knowledge management and workforce development. The session highlighted the importance of fostering leadership qualities across all levels, ensuring a diverse and inclusive approach, and preparing the next generation of leaders to navigate the complexities of the nuclear industry.

The following key points were noted during the panel discussion:

- The Role of Leadership in Knowledge Management. The panel emphasized that leadership is essential for effective knowledge management in the nuclear industry. Leaders have to empower their teams, ensuring that knowledge is not only documented but also actively shared and utilized. This approach is crucial for maintaining a knowledgeable and resilient workforce, particularly in a rapidly changing environment;

- Diversity and Inclusion in Leadership. There was a strong focus on the need for leadership to be inclusive, involving women, youth, and other underrepresented groups. The panellists shared examples of how their organizations are intentionally promoting diversity, recognizing that diverse perspectives are vital for innovation and effective decision-making in the nuclear sector;
- Balancing Management and Leadership. The discussion highlighted the distinction between management and leadership, noting that while both are necessary, they need applied appropriately depending on the context. Effective leaders need to know when to manage—focusing on processes and resources—and when to lead—inspiring and guiding their teams towards a shared vision;
- Empowering the Next Generation. The panellists discussed the importance of empowering young professionals in the nuclear industry. They shared strategies for providing young leaders with opportunities to take on responsibilities, even if it involves taking risks. Mentorship and support from experienced leaders are critical in helping young professionals learn from their mistakes and grow into effective leaders;
- Addressing Well-being and Work-Life Balance. The panel also recognized the importance of leaders attending to their own well-being and work-life balance. Leaders are encouraged to be mindful of their mental and emotional health, as this directly impacts their ability to lead effectively. Creating a supportive work environment where team members feel valued and heard was emphasized as a key aspect of leadership;
- Cultural Context in Leadership. The discussion acknowledged the influence of cultural differences on leadership styles. Leaders need to be adaptable and sensitive to the cultural contexts in which they operate, understanding that what works in one environment may not necessarily be effective in another. This flexibility is crucial for leading diverse teams in a global industry like nuclear.

In conclusion, the panel underscored the need for dynamic and inclusive leadership in the nuclear sector, capable of adapting to both current and future challenges. By fostering a culture of continuous learning, empowerment, and inclusivity, the nuclear industry can build a resilient workforce prepared to lead in a rapidly evolving landscape.

6. TECHNOLOGY: ENABLING INNOVATIONS AND KNOWLEDGE

6.1. KEYNOTE SESSION

As prepared for delivery.

V. Maugis
Head of Knowledge Capitalisation Department,
Knowledge Management Officer, ANDRA, France

The Importance of Managing the Knowledge of a Final Disposal of Nuclear Waste

Managing radioactive waste involves an imperative of preserving and passing on the memory of the facilities over several centuries. For some years now, the international community has been mobilized on this issue, with initiatives led by the IAEA or the NEA. Over and above the transmission of information from generation to generation, there is a crucial challenge, and an almost unique context, to pass on knowledge. The deep geological disposal of radioactive waste in France is meant to be operated for about one century, which means it will be run by five successive generations of workers.

So, the knowledge is to be passed on from generation to generation, and our duty is to pass on our knowledge to the generation right after us.

Since 2018, the French National Radioactive Waste Management Agency (ANDRA) is developing a specific approach dedicated to managing knowledge. Why now? We are facing two concurrent challenges. With the submission of the licensing application for the disposal, we are shifting from a phase of 30 years of design to a phase of construction: the teams who designed have to pass on the knowledge to others. And since about five years, the first generation of staff is continuing to leave. By anticipating the implementation of the project, ANDRA needs to contribute to pass on the knowledge. We have an additional pressure, in France, with the reversibility of the repository, which translates into the progressive and incremental development of the facilities. So, even if we use best available techniques as a basis, it will evolve.

Our goal at ANDRA is then to develop an approach that is in line with the standards and as operational as possible. And that is not done in an office.

Our aim is to make sure that tomorrow, managing knowledge becomes an integral part of staff activity. And my task as Knowledge Management Officer is to make sure that it becomes a routine, and they don't have to think about it anymore. That's what matters.

The first thing we should fight against is that, as we have an astronomical number of documents, holding considerable scientific and technical knowledge, we run the risk of believing that everything is written down, and that we will remember everything. But even if artificial intelligence is phenomenal, knowledge mostly resides in people's heads and in social relations. So, we should also face the fact that we miss out on most of the knowledge. The managers are missing out, the human resources officers are missing out. Eventually, the whole organisation is missing out on the knowledge.

But way beyond remedial action, knowledge management systems and strategies are integral to the functioning of the organisation. Indeed, the key knowledge to be passed on to the next generation of workers emanates directly from the very knowledge which is mobilized daily through the successive phases of the project.

Managing knowledge has to be part of employees' daily lives. And the day-to-day running of the business is luckily full of opportunities for managing knowledge. Be it the launching of a tender, or the starting of a contract, or a change of provider; a new employee arriving, an employee changing duty; the occurrence and the resolution of a problem; an audit, a control, or an appraisal... Even a very ordinary team meeting can be an excellent opportunity to discuss whether some of the required knowledge is critical or vulnerable, and to collectively consider ways of controlling it more efficiently.

There is quite a strategic dimension to managing knowledge, as it impacts safety, reliability, performance, and sustainability. So, beyond chasing the knowledge of departing staff, the strategic analysis of organizational knowledge assets may be deployed.

This is to identify and define upfront the required expertise, know-how and collective skills. To locate, within the organization, the occupations involved in the related knowledge cycles and ecosystems – be it as knowledge contributors, beneficiaries, integrators or brokers. By investigating these knowledge cycles and ecosystems, to assess the risk of loss or non-coverage of knowledge over time. To define and support control measures for the development of knowledge bases, the federating of knowledge-sharing communities, the implementation of collaborative knowledge management practices, and so on.

In essence, to support staff in accessing and controlling the knowledge they need for decision and action.

There is no one-size-fits-all solution, but a set of initiatives, for which the reasonable efforts depend directly on the nature of the considered knowledge, its target use, its criticality and vulnerability.

How does it work in practice?

At ANDRA, we are rolling out a routine approach that will be more effective in the long term and will also help to develop a cross-functional culture. For instance, in 2019 we have begun a knowledge mapping and critical analysis periodic exercise. We are building several common knowledge bases on various subjects with current employees, to homogenize practices, to share experiences, to question and to develop this knowledge. These knowledge bases help new employees to get up and running more quickly and have them contribute to knowledge management right from the start. Indeed, new employees can contribute to formalizing knowledge through their exchanges with current engineers, questioning and clarifying practices, consolidating and synthesizing existing documentation, etc. We also currently support some twenty communities of practice, covering technical and support functions and all categories of staff. These communities play a crucial role in maintaining, developing, and disseminating knowledge in a cross-functional and sustainable fashion. They can also help to onboard new employees and to support departures with the sharing of knowledge. We obviously also use internal innovation practices as well as digital collaboration modes that encourage the circulation of knowledge and cross-fertilization.

Then indeed, this approach should be integrated with the organizational functioning, to extract the most out of the synergistic combination of people, processes and technology.

We are developing it progressively, in a participative fashion, with a view to integrate daily practices gradually. Everyone needs to be involved, at every level. It's also a question of mutual concern, and managers have to encourage, support, and set an example. It's important to support the approach efficiently so that it is meaningful, operational, and based on things that "work". We start from what is possibly already in place at any scale, enlarging the scope, and focusing on strategic issues. We implement pilot actions locally, before scaling them up, integrate this into a comprehensive methodology, and finally cascade it down to every entity and activity.

And we seek to hook this with the adequate organisational devices and processes. It can be with the requirements management and with the configuration management, with the strategic planning and with the implementation roadmap, with the safety case development – obviously. Also with the integrated management system, with contracting and procurement, with professional pathways, etc. Wherever it may fit best.

We found that it works to get people to "taste", to understand by doing, and to "like" knowledge management. The beneficiaries of our actions are generally satisfied and support the approach, with some of them also keen to continue with new actions.

It takes a long time to set up this approach. But we don't start from scratch, and what takes most time is to develop habits.

Then, fortunately, we don't have key people leaving the organisation every day.

C.-L. Kim
Acting President of KEPCO International Nuclear Graduate School (KINGS), Republic of Korea

The speaker presented the development of Korean nuclear power programme through knowledge management, highlighting the vital role that KINGS plays in human resource development. He began by tracing the evolution of Korea's nuclear power programme, starting shortly after the war in 1953. Despite being a poor, war-torn country, South Korea made bold investments in nuclear energy, beginning with the TRIGA Mark II research reactor in 1959. This early commitment, including the establishment of nuclear engineering departments at Korea's top universities, laid the foundation for the country's nuclear energy capabilities. By 1978, Korea's first nuclear power plant, Kori Unit 1, began commercial operations.

Over the decades, Korea's nuclear power programme progressed in stages: from the introduction in the 1970s, to localization of technology in the 1980s, and self-reliance in the 1990s, epitomized by the development of the OPR 1000 reactor in 1995. Further innovations included the APR 1400 reactor and the ongoing development of small modular reactors (SMRs) such as SMART and iSMR. The speaker stressed that Korea's success in nuclear power stems from strong cooperation between government, academia, industry, and research institutions, with the Korea Atomic Energy Research Institute (KAERI) playing a key role. As of today, Korea operates 26 nuclear reactors and is fully self-sufficient in designing, constructing, operating, and commissioning NPPs. Korea's export of the APR 1400 reactor to the UAE in 2009 further demonstrated its global leadership in the nuclear field.

C-L. Kim then shifted focus to KINGS, which was established in 2012 to address the growing demand for highly skilled professionals in nuclear power plant engineering. Despite the existence of 16 other universities with nuclear engineering programmes, KINGS was created to specialize in nuclear power plant engineering. Located near 10 operational reactors and major nuclear facilities, KINGS provides students with unparalleled access to practical learning environments. The school offers master's and PhD programmes, with four tracks focusing on design and safety, decommissioning and spent fuel management, project management, and energy policy.

KINGS prides itself on its faculty, who are not typical university professors but seasoned professionals with an average of 20+ years of practical experience. This expertise, combined with a strong e-learning system and digital library, supports robust knowledge management and transfer at the institution. KINGS also benefits from partnerships with organizations like KHNP and IAEA, offering students mentoring programmes, simulators, and other hands-on learning opportunities. The institution has earned several international endorsements, including recognition from IAEA's International Nuclear Management Academy (INMA) and as a decommissioning collaborating centre.

Enrolment at KINGS is diverse, with 106 students currently enrolled, half of whom are international students from 21 countries. Since its founding, KINGS has graduated 552 students, with 266 international graduates from 32 countries, including 140 from Africa. KINGS provides full scholarships to international students, with the only expectation being that they return to their home countries after graduation to lead nuclear energy initiatives. Kim emphasized the school's focus on cultural exchange and communication, given its diverse student body, and the importance of preparing students for leadership roles in nuclear energy worldwide.

In conclusion, KINGS Acting President reiterated institution's mission to collaborate with domestic and international partners to foster the next generation of nuclear leaders. The school remains committed to advancing practical, hands-on education and is actively engaged in global efforts to develop nuclear human resources.

J. Sowagi
Director, Information Exchange, CANDU Owners Group Inc., Canada

Developing Nuclear Talent: A Strategic Approach to Tackling Workforce Challenges in the Nuclear Power Industry through Training

Presentation explored the proactive strategies implemented by the CANDU Owners Group in training utility staff to meet the escalating workforce demand in the nuclear sector. Focused on critical challenges such as impending retirements, hiring dynamics, and the necessity for diverse skill sets, the presentation spotlighted the innovative approaches embraced by the CANDU Owners Group.

A central theme of the discussion was the acknowledgment of the growing demand for skilled personnel and the challenges associated with retirement in the nuclear industry. The presentation elaborated on our active measures to address these challenges effectively.

Underlining the importance of catering to a diverse audience, the presentation delved into the various training methodologies employed by the CANDU Owners Group. Special emphasis was placed on addressing leadership performance weaknesses through tailored training programmes designed to enhance skills and competencies across all organizational levels.

Particular attention was given to leadership training at the supervisor and middle manager levels, encompassing both utility staff and supplier organizations. These programmes primarily target staff across the international CANDU industry spanning over six countries.

The discussion encompassed the development of succession candidates, showcasing how targeted programmes identify and nurture individuals for future leadership roles. Additionally, the incorporation of mentoring sessions within training programmes was discussed, underscoring the integral role of mentorship in professional growth and knowledge transfer within the organization.

Attendees gained valuable insights into the CANDU Owners Group's successful initiatives, illustrating a comprehensive training approach that not only meets the immediate demands of the nuclear workforce but also establishes the groundwork for a sustainable and resilient industry amidst evolving challenges.

L. O'Brien
National Nuclear Laboratory (NNL), United Kingdom

The Application of AI to Support Sustainable Nuclear Technology Development

In his presentation, speaker discussed the critical role of artificial intelligence (AI) in enhancing knowledge management within the nuclear industry. He explained that his work focuses on the implementation of AI search systems to support the management of knowledge assets, particularly within NNL's clean energy initiatives. Reflecting on his 25 years of experience, L. O'Brien emphasized that effective knowledge management can empower teams by making past information accessible, thereby increasing productivity. Conversely, poor knowledge management leads to inefficiencies and uncertainty, underscoring the need for AI to address these challenges.

The speaker outlined the structure of the NNL, which is tasked with advancing nuclear science for societal benefit. He specifically focused on one programme under the clean energy portfolio, the Advanced Modular Reactor Knowledge Capture Programme. This programme aims to accelerate the commercialization of innovative clean energy technologies, using AI to capture and manage knowledge from decades of nuclear research and development. AI's ability to search through vast, dispersed repositories of information can be a game-changer, L. O'Brien noted, particularly when the information spans decades and is spread across multiple organizations.

One of the major challenges, according to L. O'Brien, is that these large bodies of knowledge are often stored in different formats and locations. He described the use of an AI system that NNL developed, which employs an ontology structure to systematically capture, digitize, and index information – such as graphite moderator data from high-temperature gas reactors –allowing it to be searched efficiently. He explained that this process can be completed within days or weeks, depending on the complexity of the knowledge base, and has significantly reduced the time needed to access vital information compared to manual searches.

L. O'Brien also detailed how the AI system processes data, using techniques like natural language processing (NLP) and optical character recognition (OCR) to analyse and interpret text from various sources, including older documents in paper form. This capability allows the AI system to recognize concepts, locations, and parameters in the text, presenting the information in an intuitive, non-linear format that empowers researchers to quickly grasp the key points.

The system's user interface enables researchers to search for information using keywords or even natural language questions. The AI then retrieves relevant documents and auto-generates summaries, making it easier for users to interact with and share data. L. O'Brien particularly emphasized the value of concept lenses, which extract and categorize information into themes, fostering better collaboration and knowledge sharing among teams.

In conclusion, L. O'Brien stressed that AI is a powerful tool for enhancing knowledge management, offering the ability to locate and articulate critical information across diverse sources. He argued that AI not only helps in speeding up project delivery but also promotes innovation by presenting information in novel ways. However, governance and security remain essential aspects of using AI, particularly in ensuring compliance with regulations and intellectual property protections. Looking ahead, the speaker encouraged further exploration of AI's potential in knowledge management.

V. Richet
Head of Digital, ASSYSTEM STUP, France

CurieLM: A Generative AI for Smart Knowledge Management in the Nuclear Industry

V. Richet addressed the critical role of knowledge management in the nuclear industry. He began with the famous quote, "Those who cannot remember the past are condemned to repeat it," applying this to the nuclear sector, where knowledge transfer is essential. He outlined the challenges currently facing the industry, including the scarcity of resources, growing expectations for safety and transparency, and the need for a larger skilled workforce. These factors underscore the necessity for effective knowledge management, especially as legacy expertise is phased out.

The speaker described various approaches to knowledge management that have been implemented, ranging from improved document management to more advanced technologies like virtual and augmented reality. Despite these efforts, he noted that no perfect solution has yet been found. He highlighted the trade-offs between methods that are efficient but lack scalability, and those that are scalable but less efficient in the long run. He emphasized that the most critical aspect of knowledge management is making information available to the right people at the right time in a format that is easy to access.

A key part of his presentation focused on the development of Assystem's AI-based tool, CurieLM, a large language model named after Marie Curie. This model was designed to support knowledge management in the nuclear industry by providing quick and reliable answers to simple and accurate queries. V. Richet explained that AI has the advantage of being able to process large amounts of documentation more efficiently than humans, making it ideal for answering straightforward questions quickly. He presented CurieLM as a solution for addressing common knowledge queries without disrupting the production process, emphasizing its ease of use and ability to handle natural language inputs.

V. Richet provided examples of how CurieLM has been used to generate accurate answers from large corpora of documents, including those from the International Atomic Energy Agency. He demonstrated how the model can streamline the process of acquiring knowledge, which is particularly valuable for newcomers to the industry. This tool allows for faster onboarding and helps users navigate the complex jargon and acronyms that are prevalent in the nuclear field. He also noted its usefulness for integrating legacy knowledge from older projects, such as those involving long-defunct reactors.

In conclusion, V. Richet emphasized the potential of AI tools like CurieLM to transform knowledge management in the nuclear industry. He noted that while the current use case focused on answering simple queries, there are many other applications for AI, such as requirements extraction and automated reading. He encouraged the industry to leverage its vast stores of legacy data, suggesting that the barriers to utilizing these resources would decrease in the coming years. He also expressed optimism that AI would continue to enhance efficiency in the nuclear sector by reducing the time and effort needed to access critical information.

J. Reger
Technology Advisor, Former Chief Technology Officer of Fujitsu Europe (retired), Germany

What if AI is not Intelligent?

J. Reger began with a light-hearted analogy contrasting human intelligence with artificial intelligence (AI), illustrating the rapid change in societal views on technology. The speaker emphasized that while AI was pervasive, it was not yet present in every device. Many people used AI without realizing it, as it operated behind the scenes in various connected technologies. He pointed out that, although AI was foundational, it was not yet comparable to human intelligence, and its potential impact was still evolving.

J. Reger shared his personal background as a physicist and former Chief Technology Officer, explaining his current involvement in AI startups. He highlighted AI's unique position as a technology that worked through other technologies, such as quantum computing and fusion research. He noted that AI's capabilities were foundational and unprecedented, already assisting in various fields. AI, while not fully "intelligent", mimicked certain aspects of human cognition, particularly through machine learning and deep learning, which modelled aspects of the human brain with impressive results.

Despite these advancements, the speaker identified key limitations of AI, such as its lack of creativity and the opaque nature of machine learning. He introduced the emerging field of explainable AI, which aimed to clarify how AI systems reached conclusions. Another significant challenge was the energy consumption of AI systems, with current models, such as ChatGPT, requiring vast amounts of energy to operate. J. Reger emphasized the need for more efficient technologies to support AI's growth without straining resources.

The speaker also addressed issues of intellectual property (IP) and the need for AI regulation. He pointed out that AI, such as ChatGPT, consumed vast amounts of data without regard for IP rights. He stressed that although AI's rapid growth necessitated regulation, enforcing such regulations was difficult due to AI's decentralized nature. In contrast to the nuclear industry, where investments and developments were visible, AI breakthroughs could be achieved with minimal resources, making regulation complex and enforcement challenging.

AI's potential benefits to the nuclear industry, particularly in knowledge management and anomaly detection, were acknowledged. J. Reger mentioned AI's role in aiding nuclear fusion research and its potential for improving operational efficiency in nuclear facilities. However, he cautioned that AI, like all advanced technologies, carried risks, especially in high-stakes industries like nuclear energy, where extreme caution was necessary.

In conclusion, J. Reger emphasized the need to integrate AI thoughtfully, comparing its development to raising children – requiring guidance and boundaries. He saw AI as a powerful tool with the potential to shape the future of humanity. Looking ahead, he suggested that AI could help humanity evolve, potentially leading to a convergence between human intelligence and AI. While AI was not intelligent at present, J. Reger concluded that intelligence came from humans, and together with AI, there were opportunities to shape a better future.

6.3. SUMMARIES OF THE PARALLEL SESSIONS

6.3.2. Session 9.1. AI and Large Language Model Supporting HRD and NKM

Session Chair: P. Kertys, Slovakia

IAEA Session Rapporteur: A. Ganesan

The session presented six distinct approaches to leveraging AI for enhancing knowledge processes in nuclear energy. Key topics included the digitalization of design and construction planning, AI's transformative role in knowledge management for decommissioning, and document knowledge mining to extract valuable operational insights from nuclear plant records. Presentations also highlighted generative AI's ability to improve decision-making speed and accuracy, combining data-centric and human-centric approaches in nuclear plant modelling, and innovative AI applications for education and workforce development. Each presentation underscored AI's growing role in enhancing safety, efficiency, and knowledge retention across the nuclear sector.

Paper ID 347: AP1000/AP300 Design Training Model and Staffing Optimization

Speaker: A. Cruzado Lopez, Spain

The presentation begins with the introduction of advanced features, including smart innovations focusing on real cost drivers, of their new generation AP1000/AP300 NPPs and explores the benefits of digitalization in design and construction planning. Some of the key benefits highlighted are:

- Integrated digital work packages;
- State of the art integrated BIM/PLM framework;
- Digital work package interface allows full integration of construction to design;
- Procurement integration;
- Digital work package index with direct mapping to every engineering construction deliverables.

Paper ID 412: Revolutionising Nuclear Decommissioning with AI-enhanced KM: A Real-World Case Study

Speaker: D. Rourke, United Kingdom

The paper explores the potential transformative role of artificial intelligence (AI) in knowledge management (KM) for nuclear decommissioning projects. It highlights how AI-driven strategies, including natural language processing (NLP) and machine learning, streamline processes, enhance data retrieval, and codify implicit knowledge, leading to significant cost and time reductions.

Some of the benefits derived are:

- AI can automate information extraction, recognise patterns, and predict issues;
- Contextual retrieval, dynamic questioning, and secure data handling;
- Natural language processing, secure data environment, and retrieval augmented translation.

Paper ID 174: From Documents to Knowledge: Research on the Application of AI Technology in Documents Knowledge Mining of Nuclear Power Plant: A case study of Fuqing NPP in China

Speaker: J. Qui, People's Republic of China

In order to ensure the safe and stable operation, Fuqing NPP has prepared thousands of operating documents. These documents contain valuable knowledge assets that need to be deeply mined to form a knowledge base that can be effectively utilized throughout the life cycle of the NPP. This paper explains how to use AI technology to conduct in-depth research on document knowledge mining, and to extract valuable nuclear power unit operation knowledge from many nuclear power plant documents, in order to promote the continuous improvement of NPP knowledge service level. We conducted a case study of Fuqing NPP. The company developed algorithms for the structural analysis of multiple document content to achieve the conversion of documents into structured data firstly. Secondly, through the combination of named entity recognition technology and manual verification, they transformed the structured data into a directly usable knowledge base containing hundreds of thousands of knowledge nodes. Finally, using knowledge matching technology combined with specific business scenarios, they implemented knowledge retrieval, intelligent recommendations, content comparison and other service functions in the knowledge base.

Paper ID 180: The impact of Generative AI in NKM and HRD. Operational Nuclear Knowledge Enhancement and Management for new and expert users

Speaker: L. A. Piciaccia, Norway

Safety is based upon decisions; decisions are based upon knowledge. Quickly delivering relevant traceable knowledge in response to complex queries improves safety related decisions correctness, effectiveness, and speed. This paper illustrates techniques with high Technology Readiness Level in industrial and societal fields. The growing power by which AI services allow human-computer dialogue and cooperation are grounded in the ability of modern Large Language Model based systems to read, understand and elaborate textual as well as visual data. Generative AI models are first able to read inputs and interpret them properly, on a large set of contexts, domains, and scientific areas (e.g., industry, medicine, engineering, or economy). Then they enable fast, integrated data access platforms facilitating knowledge management, operating conditions departure warning or predictive inferences based on semantic analogies across very large data sets, as required by most NKM projects. The work illustrates industry level solution for knowledge gathering capabilities from raw and heterogeneous, textual, or visual data sources and extraction and indexing of the underlying concepts. The actual outcomes of the resulting framework illustrate proven ways to improve knowledge accumulation, cooperative skills and promote collaborative attitudes among large specialists and user communities.

Paper ID 364: The synergy of data-centric and human-centric approaches within application of AI technologies for NPP digital modelling

Speaker: V. Tsygoda, Russian Federation

By analysing existing research and based on own multi-year experience in creation of knowledge-intensive NPP information models in the Rosatom State Corporation Engineering Division, the goal of the paper is to provide a deeper understanding of the transformative impact of modern digital technologies, including AI, and the crucial importance of combining data-centric and human-centric approaches in knowledge capital shifting and NPP construction lifecycle management.

The presentation highlighted some of the key features in their MULTI-D application. Some of the are:

- Technical support chatbot;
- Intelligent video analytics at the site;
- Ais transformative role inn NPP construction.

The presentation tabled the current capabilities and the future possibilities using AI's transformative role in NPP construction.

Paper ID 394: Innovative Digital Approach to Nuclear fostering Education, Knowledge Management, and Workforce Development

Speaker: J. Porsmyr, Norway

The presentation highlights the transition from traditional methods to innovative digital approaches in the nuclear sector, emphasizing the importance of digital tools for knowledge management (KM), education, and workforce development. Traditional KM methods are often inefficient and lead to knowledge loss, particularly during decommissioning phases. Emerging technologies like virtual and augmented reality (VR/XR), AI-powered data analytics, and digital twins offer immersive training experiences, improve knowledge transfer, and support stakeholder engagement. The presentation underscores the benefits of personalized learning, gamification, and modern digital platforms for enhancing safety, efficiency, and collaboration. Continuous investment in these technologies is vital for the future of nuclear operations and decommissioning.

6.3.3. Session 9.2. Establishing NKM-HRD Programmes

Session Chair: J. Sowagi, Canada

IAEA Session Rapporteur: M. Ovanes

The session focused on the critical role of nuclear knowledge management (NKM) and human resource development (HRD) in ensuring sustainable nuclear energy growth, particularly in developing and newcomer countries. Key discussions highlighted the importance of fostering expertise and collaboration between academia, industry, and government to build a robust nuclear workforce. Presentations explored country-specific challenges, such as workforce gaps in Croatia and the importance of knowledge sharing in Ghana, while emphasizing the role of nuclear power in reducing greenhouse gas emissions and supporting sustainable energy. The session also covered the integration of knowledge management in radioactive waste disposal programmes and strategies for ensuring the long-term sustainability of nuclear expertise, particularly through intergenerational knowledge transfer and international collaboration. Overall, the session underscored the need for strong NKM programmes to address safety, environmental, and workforce challenges in the nuclear sector.

Paper ID325: Knowledge Management in a Developing Country: The Case Study of Ghana

Speaker: D. Nyarko, Ghana

The presentation explores how knowledge management is handled within the context of Ghana, a developing country and newcomer to nuclear power country. It discusses strategies, challenges, and best practices related to knowledge sharing, collaboration, and sustainable development. relevant given the importance of knowledge management in fostering growth and progress.

Paper ID 55: Critical Mass of Experts Required for a National Sustainable Nuclear Buildup – Croatian Case Study

Speaker: Ž. Tomšić, Croatia

The presentation discusses the history and current state of nuclear power in Croatia. It highlights the country's involvement in the Krško Nuclear Power Plant jointly own with Slovenia and Croatia's recent efforts to affirm nuclear power. The presentation also outlines the challenges faced in educating nuclear engineers in Croatia, despite the increasing interest in nuclear energy. The authors emphasize the need for a larger pool of experts for a sustainable national nuclear build-up. They conclude by stressing the importance of international collaboration in nuclear knowledge management and human resources development.

Paper ID 304: The Nuclear Knowledge Management Necessity for Sustainable Energy Development

Speaker: S. Gezer, Türkiye

The paper emphasizes the importance of nuclear knowledge management in achieving sustainable energy goals and discusses the nuclear power's role in reducing greenhouse gas emissions and delivering large-scale energy. However, challenges remain. Key considerations in addressing challenges are related to ensuring continuously high levels of nuclear safety, effective management of waste disposal and developing long-term disposal solutions for nuclear waste, as well as cost-effectiveness and investment attractiveness of nuclear power compared to other alternatives.

Paper ID 395: Lessons Learned of Knowledge Management Activities in EURAD and PREDIS

Speaker: P. Carbol, European Commission

The paper discusses the knowledge management activities in the EURAD and PREDIS programmes. These programmes, co-funded by the European Commission, aim to support EU Member States in implementing their radioactive waste disposals. The paper outlines key achievements, lessons learned, and emphasizes the importance of collaboration with knowledge providers such as IAEA and OECD/NEA. The programmes involve almost 1200 experts and over 140 students, highlighting the extensive pool of expertise available in the field of radioactive waste management.

Paper ID 438: Ensuring Sustainable Nuclear Workforce Development in Europe and Beyond

Speaker: S. Monti, European Nuclear Society

The European Nuclear Society (ENS) is addressing the critical need for a skilled workforce and effective knowledge transfer in the expanding nuclear power sector. Recognizing the importance of engaging younger generations, ENS is tackling challenges in human resources development and knowledge management across Europe and beyond. By fostering multidisciplinary networks and facilitating inter-generational exchanges, ENS aims to build a sustainable workforce equipped with the necessary skills to meet the evolving demands of the nuclear industry. This presentation highlights ENS's strategies to ensure robust workforce development and knowledge preservation.

6.3.4. Session 9.3. NKM-HRD Supporting Nuclear Waste Management and Decommissioning

Session Chair: V. Michal, IAEA

IAEA Session Rapporteur: J. Roberts

The session focused on various initiatives related to nuclear knowledge management (KM) and human resource development (HRD) in decommissioning and radioactive waste management. The EU nuclear decommissioning assistance programme and its three-phase KM strategy (2021 – 2027) were discussed, highlighting knowledge tools like an interactive platform for managing decommissioning knowledge. The integration of KM and organizational learning was emphasized, with theories such as Connectivism considered ideal for adult learning in decommissioning. Argentina's national radioactive waste management programme outlined efforts to capture knowledge from retiring staff and preserve it in cloud-based databases. The session also covered European R&D initiatives within the PREDIS project, which developed case studies to enhance KM, integrated with global information platforms. Challenges in recruitment for large decommissioning projects and the importance of competency assessments for HRD were also addressed, underscoring the need for knowledge networks and training plans.

Paper ID 255: EU Nuclear Decommissioning Knowledge Management Initiative

Speaker: A. Piagentini, European Commission

The EU Nuclear Decommissioning Knowledge Management initiative aims at gathering, screening, storing, and sharing knowledge gained by users and organisations involved in decommissioning programmes and projects in Europe. The initiative starts from creating and collecting Knowledge Products (KPs) and proceeds in their classification according to the new international decommissioning taxonomy. The final goal of DKM initiative is the creation of an interactive user-centred platform EUKLID in which KPs will be stored and made available.

Paper ID 409: How Can a Connectivism-Based Approach Enhance Adult Learning and KM in Nuclear Decommissioning?

Speaker: D. C. Invernizzi, United Kingdom

KM and Organisational Learning (OL) are strongly interconnected. Nevertheless, when discussing KM in the nuclear decommissioning industry, the focus is largely on the process related to data gathering and codification, while less attention has been posed to the process of learning from the knowledge that has been accumulated in time. The research paper leverages on adult learning theory and connectivism to address the following research question: how can a connectivism-based approach enhance adult learning and KM in nuclear decommissioning projects? The research is based on a thorough literature review and semi-structured interviews. The research findings contribute to a better understanding of how a connectivism-based approach can be used to enhance adult learning in this complex and critical sector and how the findings could shape knowledge management in nuclear decommissioning.

Paper ID 62: NKM implementation as a foundation stone in a Decommissioning Plan

Speaker: G.F. Puglia, Argentina

When discussing KM in the nuclear decommissioning industry, the focus is largely on the process related to data gathering and codification, while less attention has been posed to the process of learning from the knowledge that has been accumulated in time. This research paper leverages on adult learning theory and connectivism to address the following research question: how can a connectivism-based approach enhance adult learning and KM in nuclear decommissioning projects?

Paper ID 234: Transferring knowledge of new LILW pre-disposal practices via case studies from the Euratom PREDIS project

Speaker: E. Holt, Finland

The paper will present case studies aimed to provide guidance to the other parties on new low- and intermediate level waste (LILW) pre-disposal practices that were developed during projects. The case studies are also planned to be embedded in the IAEA Wiki and thus the KM platform is discussed from the PREDIS point-of-view.

Paper ID 210: Andra's innovative Forward-Looking HR Approach for Cigéo Construction Phase

Speaker: F. Puyade, France

In taking into consideration the specific needs of its programmes imposed by the long-life cycle of its surface disposal facilities in operation and its long-term commitments linked to Cigeo project, Andra have to address a major HR challenge, A committee tasked with assessing employment trends and resource needs was established, and implementation is engaged for 2024. This paper sets out to describe an innovative approach made in-house to anticipate the evolution of sensitive disciplines and skills some 7-10 years into the future.

Paper ID 224: The Challenges and Prospects of Human Resource Development and Nuclear Knowledge Management in the Nigerian Nuclear Regulatory Authority

Speaker: M. Akpanowo, Nigeria

This paper discusses the strategies for HRD and NKM required to fulfil the NNRA's statutory functions, the challenges of implementation, and prospects. The analysis involved identifying the regulatory functions, the tasks based on the required competencies and the associated knowledge, skills, and attitudes. The challenges of developing the requisite competencies are largely due to insufficient resources and the non-implementation of cooperative agreements with some of the NNRA partners.

6.3.5. Session 10.1. Modern Tools in HRD and NKM, Part 1

Session Chair: A. Di Trapani, Italy

IAEA Session Rapporteur: R. Kvetonova

The session discussed in detail the new and modern approaches and tools used in HRD and NKM. Many examples from digitalization and multimodal platforms have been introduced including the new opportunities for engineers, researchers, and others to test new technological solutions. The session also introduced the neuroscience research, which indicates that greater engagement, interactivity, visual aspects, and gamification improve understanding and knowledge retention. Some organizations presented examples of integration of training technologies, proprietary state-of-the-art training tools, and standardization of nuclear fundamentals programmes. Finally, good practices in use of artificial intelligence and digital transformation have been discussed too.

Paper ID 254: INVICTUS: A shared multimodal project serving human resources development

Speaker: F. Lemont, France

The INVICTUS project, developed by INSTN, aims to create a multimodal platform to address the challenges of training new employees in the nuclear sector and maintaining their skills. The platform will cover various areas such as radiation protection, waste management, and the cleanup and dismantling of nuclear installations. It will also serve as an awareness-raising tool for different audiences, including schoolchildren, the general public, and investors. The project will provide a realistic environment for training, testing equipment, and protocols for sanitation and dismantling operations. The goal is to train between 2,500 and 3,000 people per year, focusing on areas like radiation protection, interventions in nuclear environments, and management of radioactive waste. The project also aims to increase the attractiveness of the nuclear sector through open days, serious games, and thematic conferences. Finally, the INVICTUS project will offer opportunities for engineers, technicians, and researchers to test their technological solutions in environments representative of real-scale nuclear installations.

Paper ID 261: Improving Nuclear Industry Efficiency Through Neuroeducation and Neuroleadership

Speaker: P. Terry, United States of America

Neuroeducation and Neuroleadership are emerging fields that can enhance the efficiency of the nuclear industry's training investment. Neuroeducation research offers strategies that improve learner attention and cognitive processes, leading to better task performance, decision-making, and problem-solving. Neuroleadership principles, when applied, can improve training effectiveness, and facilitate better problem-solving, decision-making, collaboration, emotional regulation, and change management in the workplace. The presentation discusses these findings and their application to the unique context of the nuclear industry. The goal is to make more efficient use of the nuclear industry's training investment and substantially improve its return.

Paper ID 322: NEXA: Digitalization. Blended training approach. Standardization of curriculum

Speaker: A. Cruzado Lopez, Spain

The NEXA project, developed by Westinghouse Electric Company, aims to revolutionize training in the nuclear industry through digitalization, blended learning, and curriculum standardization. The project leverages neuroscience research, which indicates that greater engagement, interactivity, visual aspects, and gamification improve understanding and knowledge retention. NEXA considers the digital habits of new industry entrants, combining self-study with conventional learning settings. The project also contemplates outsourcing generic trainings to specialized third-party companies, providing cost savings and more advanced, standardized trainings. NEXA integrates training technologies, proprietary state-of-the-art training tools, and standardization of nuclear fundamentals programmes. It offers a complete catalogue of standard courses for various personnel and leverages Westinghouse's nuclear technology expertise for material design, development, and programme implementation.

Paper ID 425: E-learning Onboarding programme: "Iqrab W Ta'alam"

Speaker: G. Balhamar, United Arab Emirates

The Federal Authority for Nuclear Regulation (FANR) in Abu Dhabi has established an e-learning onboarding programme called "Iqrab W Ta'alam" for all new joiners. The programme provides an orientation of FANR's core business, covering topics from each department to help new joiners understand FANR's business and core functions. The programme also includes industry-related topics such as Safety Culture, Basics of Radiation and Radiation Protection, Nuclear law, Integrated Management Systems, and more. The programme aims to equip new joiners with the knowledge they need to integrate into FANR and gives them valuable insights into the departmental mandates and the importance of regulating the nuclear and non-nuclear industry in the United Arab Emirates. The programme was established in November 2022 and has been successfully completed by all new joiners since its launch.

Paper ID 360: Digital HR-services for Nuclear Industry

Speaker: O. Karmishina, Russian Federation

The paper discusses the digital transformation of HR services in the nuclear industry by Rosatom, a global company with over 370,000 employees. The company has grown significantly over the past seven years and aims to double its revenue by 2030, half of which will come from new business areas. To achieve this, Rosatom plans to hire and onboard about 375,000 people, over 20% of whom will be fresh graduates. The company is developing digital services and IT systems for effective hiring, adaptation, development, and training of specialists worldwide. The paper also discusses the systematization and centralization of HR processes, the creation of a digital ecosystem for hiring staff and personnel development, and the future plans for improving these systems using artificial intelligence. The digital solutions allow Rosatom to quickly and efficiently hire employees, provide training, and form teams for breakthrough projects.

6.3.6. Session 10.2. Lessons Learned from NKM-HRD Programmes

Session Chair: C. Chevreau, France

IAEA Session Rapporteur: H. Zhivitskaya

The session focused on five papers discussing various aspects of knowledge management and technology transfer in the nuclear sector. The first paper detailed the knowledge management model at the Nuclear and Radiation Safety Centre, highlighting its role in ensuring organizational sustainability and human resource development. The second paper examined CNEN's R&D in nuclear sciences within

Brazil's STI framework, proposing a practice-based model for evaluating Technology Transfer Offices. The third paper explored the Joint Research Centre's role in EU decision-making, emphasizing foresight activities to anticipate future trends and challenges. The fourth paper presented EHRO-N's development of a job classification system for the nuclear workforce, and the final paper described the knowledge management infrastructure established by the German Waste Management Organisation to address nuclear waste disposal challenges. Each paper emphasized the critical importance of effective knowledge management and technology transfer in advancing the nuclear field.

Paper ID 15: Knowledge Management and Human Resource Development at Nuclear and Radiation Safety Centre

Speaker: M. Simonyan, Armenia

The paper describes the knowledge management model applied at the Nuclear and Radiation Safety Centre, Technical and Scientific Support Organization to the Armenian Nuclear Regulatory Authority. It emphasizes the creation, collection, and dissemination of knowledge as fundamental assets. Employees gain valuable knowledge, experience, and skills over time, which need be documented and transferred to ensure organizational sustainability. The model addresses how institutional knowledge and expertise are created, captured, stored, shared, and integrated into work processes, including the company management system. The paper also explores the role of this model in human resource development and the challenges associated with its implementation.

Paper ID 45: Capabilities in Nuclear Sciences and Business Management: Combining Performance, Practices and Organizational Resources at the National Nuclear Energy Commission (CNEN) of Brazil for Promoting Innovation

Speaker: D. Archila, Brazil

The paper examines CNEN's extensive R&D in nuclear sciences with applications in agriculture, environment, and industry. Enhancing technology transfer and promoting innovation are key goals of Brazil's recent Science, Technology, and Innovation (STI) legal framework. This framework expanded the roles of Technology Transfer Offices (TTOs), initially established by the Innovation Law 10973-2004. TTOs now manage innovation policies, intellectual property, competitive intelligence, and technology transfer strategies. The paper proposes a practice based TTO model to evaluate CNEN's TTO performance in bridging nuclear sciences and business through sensing, seizing opportunities, cultural change management, and knowledge management. The model aims to develop TTO capabilities in intelligence, active involvement, mindset change, and knowledge management.

Paper ID 390: The Foresight Strategy Applied to the Nuclear Knowledge Management

Speaker: L. Iglesias Pérez, European Commission

The paper explores the role of the Joint Research Centre (JRC) in supporting the European Union's decision-making process. Established in 1957 and operating under the European Commission, the JRC collaborates with government agencies, research institutions, and industry partners to address challenges like climate change, energy security, and digital transformation. It emphasizes the importance of the JRC's foresight activities, which use various tools to detect early signals of future trends and technologies impacting the nuclear sector. These activities help identify knowledge gaps and adapt to evolving environments. By integrating foresight within its knowledge management framework, the JRC enhances its ability to anticipate future challenges and opportunities, providing timely support to policymakers and stakeholders.

Paper ID 386: European Human Resources Observatory in the Nuclear Field – A Status Update

Speaker: B. Eriksen, European Commission

The paper presents the work of the European Human Resources Observatory for the Nuclear Field (EHRO-N) on the development of a uniform job classification system designed to map the nuclear workforce effectively. The proposed classification framework aims to be adaptable for diverse organizational and national contexts within EU Member States, robust enough to ensure stability and comparability over time and across borders, user-friendly for human resource managers in the nuclear sector, and endorsed by major stakeholders. Additionally, the paper provides an overview of the latest survey investigating the extent of European higher nuclear education and its evolution over time.

Paper ID 320: KM Concepts and Approaches in the German Waste Management Organisation (BGE)

Speaker: G. Hoefer, Germany

The paper describes the new knowledge management (KM) infrastructure that has established based on the challenges of Nuclear Waste Disposal the German Waste Management Organisation (BGE). The infrastructure includes a connection between KM platforms and knowledge carriers to make all knowledge types available. The KM is to be implemented as a part of an BGE internal Integrated Management System.

6.3.7. Session 10.3. NKM-HRD Aspects of Research Reactors

Session Chairs: P. Chakrov and A. Shokr, IAEA

IAEA Session Rapporteur: A. Ganesan

The session focused on the critical need for effective knowledge management (KM) and human resources development (HRD) in research reactors, with each presentation highlighting national and regional experiences. The IAEA emphasized its role in supporting member states with resources and tools to enhance knowledge retention, capacity building, and reactor operations. The Nigerian presentation raised concerns about the risk of knowledge loss across key competence areas, using IAEA's methodology to assess and propose mitigations. Bangladesh shared its efforts to train and develop personnel for their TRIGA reactor but noted the pressing issue of losing experienced staff to retirement. Peru highlighted challenges in implementing KM, particularly in maintenance and operational data accessibility, and the need for stronger commitment to KM processes. RIALC's presentation stressed the importance of regional collaboration and networking, while Egypt showcased a comprehensive KM system tailored to preserve critical knowledge and improve safety, developed through IAEA guidance. Throughout the session, the discussion underscored the importance of structured KM frameworks and the urgent need to address knowledge retention as reactors age and skilled workers retire.

Presentation: Knowledge Management and Human Resources Development for Research reactors: IAEA support to member states

Speaker: P. Chakrov, and F. Naseer, IAEA

The presentation provided a brief overview of IAEA activities and tools aimed to support organizations operating research reactors and Member States developing new research reactor programmes in addressing nuclear knowledge management and human resources development issues. Capacity building and knowledge sharing tools include regular research reactor schools and training workshops, international forums and networks, Internet Reactor Laboratories and International Centres based on Research Reactors. HR Development modelling helps countries to plan human resources for their research reactor programmes at national level. Several peer review services developed by the Agency specifically for research reactors include in their scope various elements of knowledge management and HR development. Key requirements to organizations operating research reactors in the area of knowledge management and HR development are set in the IAEA safety standards.

Paper ID 25: An Assessment of the Risk of Nuclear Knowledge Loss at the Nigerian Research Reactor Facility

Speaker: U. S. Adam, Nigeria

The paper assessed the risk of knowledge loss across the various competence areas at the Nigerian research reactor organization using the IAEA methodology of knowledge loss risk management by calculating the employee total risk factor. The existing competences considered at the reactor organization are in the areas of reactor operation; neutron activation analysis (NAA); nuclear safety; nuclear security; nuclear safeguards; radiation protection; radioactive waste management; nuclear science and engineering; and traditional engineering. Based on the results of the assessment, the functions of the positions associated with each competence area were presented, and considerations to be taken into account for determining replacement needs were suggested.

Paper ID 102: National Status of Nuclear Knowledge Management and Human Recourse Development of BAEC TRIGA Research Reactor (BTRR)

Speaker: N. Jahan, Bangladesh

This paper presents the HRD, training and KM activities performed at their TRIGA Research Reactor (BTRR). The key activities highlighted are:

- Utilization of operating experiences with the compliance to regulatory requirements;
- Strengthening nuclear education, training, networking, experience exchange and research programmes;
- Different training programmes/Seminars/Workshops are organized for sharing knowledge;
- Update and preservation of critical plant documents and procedures;
- All operational and maintenance documents are stored in the control room, manager's room and in a master computer which is integrated with control room and manager's computer system;
- Cultivating an attitude of trust, openness, active collaboration, and a high degree of knowledge sharing culture within CRR working people, user groups and other stakeholders have been exercised to maintain a positive knowledge management culture;
- One challenge that needs to be addressed is potential loss of critical knowledge as many skilled and experienced reactor workers leave due to retirement and other involvements.

Paper ID 396: Knowledge Management at Nuclear Reactor División in RP10

Speaker: A. Zúñiga, Peru

This paper discusses the implementation of Knowledge Management (KM) at the nuclear reactor division of the Peruvian Institute of Nuclear Energy (IPEN). KM is crucial in organizations where safety is paramount in all activities related to nuclear technology.

Current State of Reactors: The division is responsible for the operation of the RPO (1 watt) and RP10 (10 megawatts) nuclear reactors. The RPO is currently in prolonged shutdown (5 years), and the RP10 is underutilized (no more than 15% of its capacity).

Methodology and Action Plan: The method involves identifying the contingencies of the division which influence the bases of KM and the solutions and processes. The KM survey has been developed using the IAEA guide. Problems and responsible areas have been identified, and solutions are proposed in the form of an action plan that would extend to one year, contributing to a more efficient response of the nuclear reactors.

Some of the KM challenges highlighted are:

- No strong decision in favour of implementing KM;
- The most challenging area is maintenance;
- Information is not easily accessible in operations department.

Presentation: Regional Network of Research Reactors and related institutions in Latin America (RIALC)

Speaker: A. Zúñiga, Peru

The presentation focused on the Regional Network of Research Reactors in Latin America and the Caribbean (RIALC). The RIALC network, launched in 2023 and coordinated by Peru, aims to address the increasing demand for nuclear research reactor products and services in the region. The presentation highlighted the involvement of nine countries – Argentina, Bolivia, Brazil, Chile, Colombia, Cuba, Jamaica, Mexico, and Peru – and emphasized the reactors' contributions to various sectors like medicine, agriculture, and industry. Key proposals included fostering human capital development through regional training, enhancing research collaborations on topics such as health, climate change, and agriculture, and utilizing existing nuclear infrastructure more efficiently. The network also focuses on public engagement, publishing educational materials and organizing regional events to promote nuclear science and technology.

Paper ID 58: Nuclear Knowledge Management System for Egypt Research Reactors

Speaker: S. El-Morshedy, Egypt

This study addresses a knowledge management system for Egypt research reactors. Egypt has two research reactors namely ETRR-1 and ETRR-2. ETRR-1 went critical for the first time in fall 1961 reactor and in extended shutdown state since April 2010. ETRR-2 achieved initial criticality On November 1997, and is used mainly for radioisotope production. The present nuclear management system is designed based on the elements provided by the IAEA Knowledge Management Assist Visit (KMAV). These elements include policy and strategy, human resource planning and processes for knowledge management, training and human performance improvement, document management, technical IT solutions, tacit knowledge capture, knowledge management culture and the external collaboration. This knowledge management system is vital for preserving the accumulated knowledge and experience and making safety the top priority. It also positively influences employee's attitudes and behaviours regarding safety.

6.4. MODERATED PANEL DISCUSSION ON NEW TECHNOLOGY

Panel discussion was moderated by J. Reger, Germany.

Panellist:

V. Richet, Head of Digital, ASSYSTEM STUP, France

P. Terry, Manager, Skills and Proficiency, Westinghouse Electric Company LLC, USA

N. Ngoy Kubelwa, Nuclear Engineer (Instrumentation and Control Systems), IAEA

L. O'Brien, Laboratory Fellow, National Nuclear Laboratory, United Kingdom

P. Kertys, Head of Data Science, Slovenské Elektrárne, Slovakia

Panel discussion on New Technology was titled "New Technologies, People, and Organizations: Who Shapes Whom?", it explored the dynamic relationship between emerging technologies, particularly artificial intelligence (AI), and the nuclear industry. The session aimed to explore how these technologies are transforming the nuclear sector, influencing organizational structures, and reshaping the workforce.

The following key points were noted during the panel discussion:

- Integration of AI in the Nuclear Industry. The panellists collectively acknowledged AI as a transformative force in the nuclear industry, fundamentally different from previous technological advancements. AI's capabilities in areas such as predictive analytics, skills acquisition, maintenance support, and diagnostics were highlighted as significant value-adds. By automating complex tasks and enhancing decision-making processes, AI offers the potential to streamline operations and improve efficiency across the nuclear sector. However, the panel emphasized that AI should be viewed as a tool to support, not replace, human expertise, with its role being to complement human judgment in critical decision-making processes;
- Challenges of AI Adoption. Adopting AI in the nuclear industry presents several challenges, particularly in integrating new technologies with legacy systems that have been in place for decades. The panellists discussed the difficulties of ensuring that AI tools can work seamlessly with existing data and infrastructures without requiring costly overhauls. Additionally, there is a need to address workforce apprehension towards AI, which can stem from a lack of understanding or fear of obsolescence. Clear communication, consistent terminology, and thorough pre-deployment analysis are essential to overcoming these challenges and ensuring a smooth transition to AI-enhanced operations;
- Impact on Workforce Skills and Training. AI's integration into the nuclear industry is expected to significantly alter the skill sets required of the workforce. While AI can take over routine and repetitive tasks, this shift necessitates comprehensive retraining programmes to ensure that employees are equipped to handle more strategic, creative, and complex aspects of their roles. The panel stressed the importance of maintaining critical thinking and problem-solving abilities among employees, even as AI assumes more operational tasks. Training needs to be tailored to different levels within the organization, ensuring that all staff can effectively interact with AI and leverage its capabilities;
- The Role of Leadership in the AI Era. Leadership within the nuclear industry will undergo significant changes as AI becomes more integrated into operations. The panel discussed how AI can democratize access to information and enhance decision-making, but it also shifts the role of leaders from traditional management tasks to more strategic oversight and creative problem-solving. Leaders will need to adapt to new ways of working, guiding their teams through the technological changes while ensuring that AI is used responsibly and effectively.

The need for leaders to be well-versed in AI technologies themselves was also emphasized, as their influence will be critical in navigating the industry through these advancements;

- Knowledge Management and AI. AI has the potential to revolutionize knowledge management in the nuclear industry by automating the capture, storage, and dissemination of information. This could be particularly valuable in preserving institutional knowledge and aiding the onboarding of new employees. However, the panel cautioned against over-reliance on AI for knowledge management, noting that it could disconnect the emotional and social aspects of learning that are vital for effective decision-making. Ensuring that AI systems are fed with accurate and complete data is crucial, as any flaws in the data could lead to significant errors in knowledge management;

- Safety and Security Considerations. The panel highlighted the importance of managing the integration of AI into safety-critical areas with extreme caution. While AI offers significant benefits, such as improved efficiency and enhanced decision-making, it also introduces new risks, particularly in the realm of safety and security. The IAEA and other organizations are working to develop standards and guidelines to ensure that AI is implemented safely in the nuclear sector. The panellists underscored the need for robust validation and verification processes to ensure that AI applications meet the stringent safety standards required in this highly regulated industry.

The panel discussion provided a comprehensive overview of the opportunities and challenges presented by AI in the nuclear industry. The panellists emphasized that while AI has the potential to greatly enhance efficiency, decision-making, and knowledge management, its implementation needs to be carefully managed to avoid unintended consequences. The role of leadership in guiding this transition will be crucial, as leaders will need to balance the benefits of AI with the need to maintain safety, security, and human expertise.

The panel concluded that AI will play a significant role in the future of the nuclear industry, but it will require a concerted effort from all stakeholders to ensure that it is used effectively and responsibly. The session highlighted the importance of preparing the workforce for these changes through comprehensive training, clear communication, and robust leadership. As AI continues to evolve, the nuclear industry has to remain proactive in adapting to these technologies while maintaining the highest standards of safety and security.

7. ALLIANCES: ENGAGING YOUTH THROUGH GLOBAL COLLABORATION

7.1. KEYNOTE SESSION

As prepared for delivery.

E. Pule

President of the Conference, HR Executive of Eskom Holdings SOC Ltd., South Africa

Good morning, ladies and gentlemen, esteemed delegates, and future pioneers of the nuclear energy industry.

I am Elsie Pule, the Group Executive of Human Resources at Eskom Holdings in South Africa. Before I get into the topic for today, I'd like to share an overview of Eskom and how we utilise nuclear energy. Eskom is a 100% state owned electricity utility that supplies approx. 90% of South Africa's electricity. We have over 30 power stations with a total of 47TW capacity. One of our power stations is the Koeberg Nuclear power station, which is Africa's first and only nuclear power plant, making up about 4% of Eskom's energy mix. We have been operating Koeberg for 40 years and have recently completed projects to safely extend the life of the plant.

We as Eskom, are faced with the same challenges as most of you are i.e. Attracting and retaining younger generations of talent in the nuclear sector. It is with great honour that I stand before you today in my capacity as President of this year's conference to address you on this subject.

Importance of the topic

We have observed a global decline in the number of young people studying towards careers in the sector. Global Energy Talent Indices, which survey people working in the nuclear sector across 166 countries show that only 30% of respondents were between the ages of 18 and 34, compared to 36% who were over 55. One of the major challenges we face is the preservation and transfer of knowledge from experienced professionals to the subsequent generations. When a significant portion of the nuclear workforce approaches retirement, there is a risk of losing critical expertise and institutional memory. Effective knowledge management is critical in ensuring continuity and preventing knowledge gaps over time.

A survey by the Institute of Mechanical Engineers, found that, among young people, there is a general scepticism towards nuclear power and an unawareness of its role as a low carbon source of energy. According to the survey, young people are concerned about the safety of nuclear energy, especially when it comes to the management of nuclear waste. Investing in education and providing hands-on experience through internships and mentorship programmes are fundamental steps towards building a knowledgeable and skilled workforce and allaying these fears and misconceptions.

The nuclear sector has been a cornerstone of scientific advancement and a key contributor to global energy security. However, the safety, security, and sustainability of this field hinges on the effective pipelining of talent with fresh perspectives, innovative ideas, and dynamic energy that young minds bring to the table. Our mission, therefore, is not only to pass the torch but to ensure that the flame of curiosity and the zeal for discovery burns brighter in the hearts of the youth.

We are at a pivotal moment where the convergence of technology, environmental consciousness, and global collaboration creates an unparalleled opportunity for the nuclear industry. To harness this potential, we have to create pathways that are not just accessible but also appealing to the younger generation. We are also seeing a critical disconnect persisting between professional and entry level training with business needs, which is why we often lose talent early on in their careers.

Current initiatives

To attract young talent into the nuclear sector, several initiatives have already been put in place by organisations in the sector. This needs to be strengthened through the development of further programmes, sourcing of funding and harnessing the collaborative opportunities between countries and organisations. Some of the initiatives we have seen in place over the last few years include:

Inclusion – young talent at the forefront

Understanding what motivates existing young talent to join the sector and leveraging those insights. You will see at this year's conference we have the ProSTEM+ programme which is a Challenge that invites young from utilities, universities and training centres, secondary and vocational schools, research and design organizations, regulatory bodies, manufacturing and service companies, as well as young students and professionals from different areas related to the energy sector, to propose their innovative ideas and outreach projects on attracting and developing the new generation of workforce in STEM related specialities. The UN has also been placing young people at the forefront by encouraging those in internships, ambassador programmes or scholarships to share their personal experiences with other young people.

Expos, workshops, and conferences

The IAEA and other organizations in the sector, including Eskom, host and facilitate conferences and workshops in order to attract young people into nuclear or STEM fields. At Eskom, for example, we host an annual EXPO for Young Scientists at primary and secondary education level. Targeting young people before they begin their tertiary education.

Career roadmaps

The International Atomic Energy Agency (IAEA) emphasizes the importance of providing clear career roadmaps for young learners and professionals, which is crucial for their career advancement and retention in the industry. Companies can do this locally.

Internships and scholarships

We have seen young people enter the sector on scholarships and internships, which allow them to enter the working world and gain exposure to the industry. The main sources of motivation here are the funding and the length of the programmes, that are made available to young people who often cannot afford to study further or are looking for practical experience and employment. Organisations need to create more of these opportunities, attracting young people from rural areas, to widen the talent pool.

Our approach as Eskom and South Africa

'Training facilities'.Eskom's has an established Academy of Learning whose sole purpose is to deliver technical, functional and leadership training in the energy industry. We aim to close the skills gap across South Africa, and thereafter expand through partnerships across Africa and Globally. Together with this centralised learning academy, we have also established localised training centres e.g. Koeberg Nuclear Power Station's Training Facility which stands as a testament to this commitment by housing state-of-the-art training facilities for nuclear and renewables proficiencies for employees and communities in South Africa. On a yearly basis Koeberg also embarks on an Initial License Training (ILT) programme to help individuals qualify as Reactor Operators as this licenced skill is scarce within the nuclear industry in our country.

'Partnerships'. Eskom has also established several strategic partnerships and twinning agreements. Most recently, we have partnered with. Rosatom for Nuclear Scholarships in Russia and we continue to encourage our staff to gain international exposure on assignments with organisations such as WANO and INPO. We have partnered with countries such as China who have offered Masters in Nuclear Engineering scholarships at Tshinguoa University, Kepco International Nuclear Graduate School in Korea (KINGS) is providing similar full funded scholarships and capacity development programmes. Earlier this year, we signed an MOU with Russia's ROSATOM to source a further 30 scholarships for nuclear related studies. These are some of the jointly developed programmes which we can implement to drive this shared vision for the development of young people.

'Crowdsourcing'. Our partnership with ROSATOM includes exploring further global collaborations through initiatives such as global crowdsourcing which we aim to use to gather skills or encourage public participation to solve specific technical challenges in the nuclear field. We adopted a similar approach locally at Eskom which helped us plug some of our critical skills gaps. Crowdsourcing can be used to connect experienced nuclear professionals with students and young engineers and can be leveraged to increase funding for scholarships for students seeking education in nuclear sciences.

'Project 100'. During 2016, we made a conscious decision to proactively create a pipeline of nuclear skills which are scarce in South Africa. We then went out on a dedicated recruitment programme called "Project 100". This resulted in the appointment of 100 young trainees for our nuclear power plant operating and maintenance pipeline. The primary purpose of these appointments was to undertake an accelerated Skills Development Programme to address the future skills needs of Koeberg Power Plant to build a sustainable nuclear skills pipeline whilst simultaneously addressing shortcomings in the national demographic profile of staff in the Operating Department. The learners are recruited from institutions of higher learning and undergo 2-3 years rigorous training. The completion of the programme has resulted in opportunities for permanent jobs for the individuals who started as trainees. It was also important for us to address issues of diversity, by ensuring this pipeline of 100 trainees was made up of 40% of young women. Most of which had not previously considered a career in nuclear. These young people are now participating in the industry and some of them recently attended the AtomEXPO in Russia earlier this year.

'National initiatives'. South Africa has been proactive in developing nuclear skills through a variety of initiatives and partnerships. The Nuclear Energy Corporation of SA (NECSA) Learning Academy is a prime example, offering a highly accredited nuclear skills training centre that meets the needs of clients both nationally and internationally. With a mission to equip participants with professional and technical skills, the academy provides training that adheres to national and international quality requirements. The Department of Minerals and Energy supports skills development with bursaries, internships, and the Contract Energy Officer programme, while industry entities offer similar opportunities. Partnerships with the South African Institute of Welders and the International Atomic Energy Agency further enhance the training landscape. Additionally, joint ventures like ARECSA, established between Ariva and Necsa, focus on skills development funding. The Nuclear Energy Corporation of South Africa (Necsa) has also established the Nuclear Skills Development Centre, and there are initiatives to absorb top scientists and engineers from projects like the Pebble Bed Modular Reactor. The National Nuclear Regulator (NNR) contributes with its own set of programmes, including the Talent Management Forum and the Women in Nuclear schools outreach project. These comprehensive efforts ensure that South Africa continues to build a robust nuclear skills base, essential for the country's energy sector.

Next steps or action required

Encouraging young people to pursue careers in nuclear science and technology can help countries maximize the benefits of atomic energy for development; however, many countries, particularly developing countries, often face challenges with youth engagement. There are a number of things we can do to solve this challenge.

'Global and local collaboration'. Earlier today we had a panel discussion about global collaboration. As defined by Sustainable Development Goal 17 (SDG17), strengthening global partnerships is integral to sustainable development and meeting the Global Goals. Collaboration and partnerships play a pivotal role in equipping young people with the necessary skills to thrive in the modern workforce. Countries can create coalitions that address the attraction, development, and mobilization of a nuclear workforce. This includes targeted skills interventions at schools and the development of apprenticeship programmes tailored to keep skills within the sector. It is also important that once we attract young talent, that we nurture and retain them within the nuclear industry. To help young people to engage and initiate programmes such as the United Nations Nuclear Young Generation (UNNYG), outreach and educational programmes supported by the IAEA and groups like the UNNYG are essential.

'Mentorship, support and exposure'. To avoid repeating safety issues from the past, companies can invest now to ensure the proper transfer of knowledge. Networking and mentorship programmes play a dual role in transferring knowledge and offering the career progression young people want when entering a job. The IAEA hosts events such as the Youth in Nuclear: Engaging the Next Generation of Leaders' event that took place previously. Sessions such as these encourage learning and growth from young people and offer tenured or older talent to provide guidance and support where necessary. We have an important role as policymakers, established professionals, and organizations in raising the interest of young people in the nuclear field and providing opportunities to build knowledge, skills and networks.

'Overcoming hurdles for global mobility'. The Global Energy Talent report for 2024 states that "The nuclear sector is the odd one out when it comes to energy industry mobility. That may be because nuclear power plants, more than others, tend to create long-term communities around them, so that workers may feel deeper roots. Security clearance can also be a significant administrative hurdle when moving between countries. Companies and countries rightly take nuclear security very seriously. Coupled with lacklustre performance on salary growth in the sector, a lower interest from new talent is observed". These are some practical challenges we need to overcome through well-thought-out policies and procedures between countries and organisations.

'Digitisation and technology'. The digital revolution in nuclear safety represents an opportunity for young people. Intelligence (AI) and Machine Learning (ML) can be used to personalise learning experiences and improve learning outcomes. AI tutors and chatbots can also support learners outside traditional settings. The implementation of remote Augmented Reality (AR) and Virtual Reality (VR) can make practical learning safe in the nuclear industry. We can also incorporate Gamification or game design elements into the learning process to enhance engagement and motivation, making learning more interactive and enjoyable from a very young age. There is an opportunity for ambitious young nuclear professionals to step into this gap and provide the leadership and innovation that the sector will need. Technical skills will always be in demand, but tomorrow's most valuable professionals will be those who can bridge the technical and 'human' side, especially as AI becomes more widespread. The nuclear industry is also experiencing a renaissance with the rise of Small Modular Reactors (SMRs), which are attracting numerous start-ups and young talents using digital tools. This is part of a broader trend where the integration of digital technologies into the nuclear sector is seen as a key opportunity to draw in a younger workforce.

'Communication and awareness'. To attract younger generations, we need to communicate the profound impact that nuclear science has on everyday life, from medicine to agriculture, and from energy to space exploration. The narrative around nuclear energy need to evolve to resonate with the values and aspirations of the youth. Sustainability, safety, and the use of advanced technologies such as artificial intelligence and robotics in nuclear applications are topics that spark interest and can draw young talent towards our sector. There are a lot of activities that are nuclear related, but there are a lot of missed opportunities because we are not communicating and exposing the young to nuclear science and technology.

'Education and on-job experience'. We have to also foster an environment that encourages creativity, supports risk-taking, and celebrates innovation. This means not only equipping young professionals with the technical know-how but also nurturing their leadership and problem-solving abilities. It is important to also encourage non-STEM or non-nuclear skills to enter the industry. It is important to stress that the nuclear field is a very diverse one, meaning that we will not only need skilled engineers and scientists but also other professionals such as economists, human resource managers, communication specialists etc.

'Diversity and Inclusion'. Diversity and inclusion are crucial for young people as they foster a sense of belonging and respect for differences. Embracing diversity helps to diminish discrimination and promotes empathy towards various cultures and traditions. For young individuals, experiencing diversity can lead to innovative thinking and problem-solving skills by breaking down barriers and encouraging a global perspective. Inclusion in this context means providing equal opportunities for all, regardless of background, which is essential for progress and the pursuit of freedom. Inclusion and diversity have to be therefore at the forefront of our agenda. A diverse workforce is a resilient workforce, capable of thinking outside the conventional paradigms and driving progress. We should actively work to break down barriers and create an inclusive culture that welcomes young professionals from all walks of life. This is especially important as we aim to encourage more women to join the field as well.

'Gender equality'. We have to encourage young girls and women to pursue science and technology. Though more women are entering these technical sectors, including the nuclear field, more effort is needed to involve and support young women and girls. The Global Women in Nuclear Organisation for example, have strategic initiatives such as the Global Young Generation Group, in place for attracting and uniting young people within their community, thereby bridging generational gaps, and benefitting from the experienced women and senior leaders that can be role models and mentors, facilitating opportunities for professional development for young women and girls.

'Think about Africa'. Although in the past, nuclear energy has been a minor contributor in African countries, many especially in the north, are scaling up the use of nuclear science toward development, and some are considering the introduction of nuclear power programmes. With over 60% of the population in Africa under 25, the continent has the world's largest youth population relative to its size. This an untapped talent pool waiting to be explored. We are seeing young people raise their hands. The African Young Generation in Nuclear (AYGN) organisation is a youth-led, non-profit and non-partisan organization that brings together national networks of young professionals in nuclear and other related fields. We need to ask; how can we partner with them and other young people in the sector?

'Young people, be fearless'. And to the young people out there. My advice is that youth should not be afraid to pursue what they can conceptualize in their minds in pursuit of their dreams. And to create platforms to induce knowledge sharing to embrace nuclear science and technology. Apply for that job! Don't be afraid to apply for positions where you may feel you aren't 100% suitable for. Even if you are rejected, keep applying for the roles that interest you. Companies do not expect you to know everything. You will close the gap on the job provided you take the opportunity to learn. Perseverance is important. Leverage social media to network and find opportunities. Get help with references, letters of recommendation and CV writing. Write to companies to learn more about how the recruitment process works or what the company is like. Join live sessions and webinars to learn more about the company's programmes and projects. This gives you an advantage in interviews. Think beyond your borders and apply to other countries Take action!

Conclusion

There is only one way to prepare for the challenges of the future, it is engaging youth. As we look to the future, organisations need to commit to not only sharing knowledge but also to listening. Listening to the aspirations of young generations, understanding their vision for the future, and collaborating with them to make that vision a reality. This a long-term investment, and we need to expose the young generation to the benefits of science and technology. Youth is our future. We need to encourage them and create conducive opportunities for them to take the lead in the development of applications of nuclear science and technology. Together, we can ensure that the nuclear sector continues to thrive, powered by the passion and ingenuity of the brightest young minds from around the globe. Let's continue to build on the momentum generated at today's event and underscored the IAEA's ongoing commitment to supporting countries in Africa and worldwide in engaging the next generation of leaders in nuclear science and technology.

A. Aszodi
Professor and Dean at Budapest University of Technology and Economics

Can the nuclear industry really attract young people

A. Aszodi provided an overview of the challenges facing the nuclear industry and engineering education, particularly in Hungary. He highlighted the growing pressure on the nuclear industry due to global energy policy objectives, particularly the ambitious target of tripling nuclear capacity by 2050. The keynote speaker emphasized that this goal puts immense pressure on the sector, as it requires both the extension of existing reactor lifespans and the construction of new reactors, including large light-water reactors and, potentially, small modular reactors (SMRs). This expansion, while crucial for achieving carbon-free energy goals, presents significant staffing challenges, especially in attracting skilled engineers to rural locations where most nuclear facilities are situated.

One of the key issues raised was the difficulty in attracting young engineers to the nuclear sector. Many talented students are drawn to emerging fields like fusion and SMRs, which are perceived as more modern and innovative. Additionally, there is strong competition for talent across various high-tech sectors. The nuclear industry, therefore, needs to engage students early in their education, offering research opportunities and contracts before they are drawn into other industries. Waiting until after graduation to recruit students is often too late.

The keynote speaker also addressed challenges in the Hungarian education system, particularly the recent reduction in core mathematical content in secondary schools. These changes have left students less prepared for university-level engineering courses. Without fundamental knowledge of key mathematical concepts, students struggle in engineering programmes. In response, the university is implementing a system to assess and improve students' math skills upon entry, helping to bridge these gaps.

Additionally, the high dropout rate in engineering programmes, with about 35% of students leaving after the first year, was a major concern. Today's students have different knowledge profiles compared to previous generations – they excel in language skills, communication, and technology but may struggle with traditional academic content. Universities need to adapt their teaching methods to better engage this generation and reduce dropout rates.

In conclusion, A. Aszodi called for stronger cooperation between governments, industries, and educational institutions to address the challenges facing both the nuclear sector and STEM education. He emphasized the need for early intervention in primary and secondary schools to prepare students for future careers in STEM fields. By engaging students early and providing them with the necessary skills and motivation, the nuclear industry and related fields can better secure the talent needed to meet future demands.

K. Yamashita
Project Advisor, JAIF International Cooperation Centre (JICC), Japan

The keynote speaker discussed Japan's experience with Nuclear-Human Resource Development (N-HRD) in the nuclear sector. He began by acknowledging the impact of the Fukushima Dai-Ichi nuclear accident. Following the disaster, Japan's nuclear policy underwent significant changes, resulting in a renewed focus on nuclear energy, particularly for energy security and decarbonization. As nuclear power plants were restarted after years of suspension, there was an increasing demand for skilled engineers and human resource development in the nuclear sector became a critical challenge.

He highlighted that a significant problem Japan faced was attracting young engineers to the nuclear field. Most university students preferred other industries like automotive engineering, with only a small percentage showing interest in nuclear engineering. He noted that this was one of the key challenges addressed by Japan N-HRD network. In addition to attracting students, Japan also needed to transfer knowledge from senior engineers, who were nearing retirement, to the younger generation. This required an organized network to facilitate effective knowledge transfer and capacity building.

He introduced the Japan Nuclear HRD Network, which involved collaboration between government, universities, and industries. The network, supported by the government and key nuclear organizations, served as a platform for stakeholders to work together on developing human resources. K. Yamashita described how the network fostered cooperation among different sectors, including universities, research institutions, and power companies. The network's activities included training programmes, facility tours, and practical sessions that allowed students and young engineers to experience nuclear technology firsthand.

The speaker emphasized the importance of early education in nuclear and radiation science, noting that this education began at the elementary school level in Japan. This foundational education was designed to enhance proper understanding of nuclear energy and radiation. He also mentioned that Japanese research institutions played a crucial role in providing training and hands-on experiences for students, which helped bridge the gap between theoretical knowledge and practical application. Collaboration between universities ensured that resources, including expert lecturers, could be shared across institutions.

He further elaborated on Japan's international cooperation in HRD, particularly with the IAEA. The Japan-IAEA Nuclear Energy Management School was one of the key initiatives aimed at developing leadership and management skills in the nuclear field. Through these programmes, Japanese and international students alike benefited from enhanced learning opportunities and professional exchanges. K. Yamashita also mentioned the importance of leadership training in fostering the next generation of nuclear experts who could navigate the complexities of the global nuclear industry.

In conclusion, K. Yamashita stressed the importance of collaboration among government, industry, and academia in addressing Japan's HRD challenges. He recommended that IAEA member states establish national Nuclear HRD networks, similar to Japan's model, to support capacity building in their own nuclear sectors. Through such networks, countries could address complex challenges that individual organizations could not solve alone. He also touched on the importance of public understanding, noting that networks could help enhance public engagement and trust in nuclear energy.

7.2. PLENARY SESSION

K. Madden
President of the International Youth Nuclear Congress (IYNC),
Safeguards Evaluator, Department of Safeguards, IAEA

The speaker emphasized the mission and activities of the International Youth Nuclear Congress (IYNC), an organization dedicated to empowering and developing young professionals in the nuclear industry. IYNC operated on three key pillars: capacity building, leadership development, and youth participation in decision-making processes. With a network of over 100,000 young professionals and students across 48 countries, IYNC's mission was to promote the peaceful uses of nuclear energy and encourage youth involvement in global nuclear initiatives.

One of the major focuses of IYNC was to integrate youth voices into decision-making processes within the nuclear sector. The organization ran several programmes, including the biannual IYNC conference, which fostered global networking and knowledge exchange among young professionals. K. Madden highlighted the significance of engaging beyond the nuclear sector, particularly on social issues like climate change, which was a key driver of youth interest in nuclear energy. IYNC collaborated with various organizations, including the UN's YUNGO group, to increase nuclear literacy and promote nuclear energy as a solution to climate change.

She also discussed IYNC's World Young Generation in Nuclear Thermometer Project, which aimed to understand the motivations and challenges faced by young professionals entering the nuclear industry. The project revealed that addressing climate change was a primary motivator for youth involvement. IYNC focused on creating synergies between nuclear and other sectors to drive innovation and solve global challenges, particularly through initiatives like the Climate Talks, which encouraged dialogue between the nuclear industry and broader communities.

To further promote innovation and youth participation, IYNC ran contests such as Innovation for Nuclear, which allowed young professionals to develop and present innovative ideas to address global challenges. The contest was part of a broader effort to align the nuclear industry's messaging with the values and aspirations of the younger generation, who were increasingly focused on social impact rather than just high-paying jobs. K. Madden stressed that the nuclear industry needed to attract a diverse range of professionals, not just nuclear engineers, to meet future challenges and expand the workforce.

Another major theme of Madden's presentation was the importance of intergenerational collaboration in decision-making. She argued that nuclear projects, due to their long-term nature, required diverse perspectives from different generations, genders, and geographical regions to ensure inclusive and sustainable decisions. IYNC had launched the Leaders for Nuclear programme, which focused on developing leadership and communication skills among young professionals, fostering intergenerational dialogue, and enabling youth to contribute meaningfully to the global nuclear discourse.

K. Madden concluded by inviting attendees to participate in IYNC 2024, an event designed to provide young professionals with technical and leadership skills, as well as opportunities for global networking and cultural exchange. Through workshops, mentoring programmes, and technical tours, IYNC 2024 aimed to continue fostering intergenerational collaboration and knowledge transfer within the nuclear industry.

As prepared for delivery.

E. Rakhmankina
Deputy Director General for Human Resources Management and
Organizational Development, Rosatom, Russian Federation

1. Human-centric approach for attracting young engineers.

2. Today, Rosatom is a leader in tradition fuel business and the production of enriched uranium product and nuclear fuel:

- Every 6th power unit in the world fuelled by Rosatom;
- Rosatom has 17% of the global nuclear fuel market;
- More than 10 countries are supplied with fuel and its components for research reactors.

Also are we implementing a large number of high-tech initiatives that go beyond nuclear power.

3. Growing production volumes in all areas of Rosatom's business require hiring more than 25,000 engineers by 2030. In the recruitment plan, Rosatom rely not only on qualified experienced workers and engineers, but also, first of all, on young people.

From 2011 to 2021, the number of 20–24-year-olds in Russia was almost halved as a consequence of the lower birth rate in the late 1990s.

Given the explosive growth of science and production in Russia in recent years, new jobs are being created in a multi-fold larger volume than in previous periods and competition for personnel is becoming more urgent.

In response to the challenge of a multi-fold increase in the need for engineering and working talents due to the growth of production volumes and the development of new technologies, Rosatom adapted the processes of attracting, training, and career development of young people, taking into account generational characteristics. The updated approaches are based on the principles of human-centricity and customer-centricity.

4. To form a portrait of the new generation of young nuclear industry engineers, sociological studies were conducted.

So, there are no differences between young and matured engineers' values in:

- High level of desire to learn and improve their skills;
- Safety culture;
- Need for a team, professional and social communities;
- Desire to be involved in a cause that is significant for the country and the world;
- Company's support for participation in volunteering, professional associations, sports and creative activities.

But some zones of conflict between different generations of engineers were diagnosed in:

- Career;
- Feedback on quality of work, support and recognition;
- The essence of the work;
- Work-life balance;

- Perception of information.

We found that the young want quick and clearly defined career plan, they like new tasks, feedback and recognition are vital for them, and the work-life balance is required. Corporate souvenirs and contests are important for them.

So, we have updated our system of for attracting, training, and immersing new engineers into the culture of a nuclear company according to these features of the new generation.

5. The key principles of fine-tuning tools for effective corporate integration of young engineers are:

The approach of HR-marketing and customer-centricity: personal offers or solutions stemmed from the needs of a particular candidate or employee, frequent face-to-face contacts, involvement and building the loyalty to the company's values, collecting feedback, continuously improving the company's internal processes to provide an employee with a comfortable self-perception within a team and understanding of their own value.

Human-centricity: a set of opportunities for each potential employee to be successful in different roles throughout one's life.

Ecosystem mode: not a set of non-connected activities, but a set of integrated measures to engage and form a stable inner interest and career preferences, starting from a young age.

Gamification and educational content that generates and supports interest in the profession and industry challenges.

6. Our main tools for attracting and engaging young engineers:

Ecosystem "Kindergarten – School-College – University – Work-Teaching"

The game format of interaction for children and content meetings with parents, competitions and engineering shifts for schoolchildren allow to consciously approach the choice of a career path.

Comprehensive information campaigns in universities.

Branding and technical equipment of classrooms and recreation areas, placement of career guidance information in public spaces as well as early career guidance for students led by active successful engineers and workers.

Wide range of professional competitions and challenges, technical tours and quizzes help schoolchildren and students to get attracted to engineering activities and acquainted with Rosatom's corporate culture and values.

Industrial integration with universities is implemented at the country level through the development of Advanced Engineering Schools, whose key goal is to train highly qualified engineers of a new generation in high-demand areas.

Targeted training agreements and preliminary agreements on future employment help students to develop an understanding of their own prospects, value, and professional relevance.

Industry-specific training centres of Rosatom engineering competencies based on enterprises and specialized colleges, one of the key goals of which is to educate and train students from different universities for the key required professions of Rosatom on the example of real-world tasks. Training in Rosatom's competence centres is often followed by the participation of students and schoolchildren in the AtomSkills competition.

Special events for parents and teachers

Rosatom Teacher Programme for transferring critical knowledge to young professional teachers and engineers.

7. Our key tools to engage young employees are:

Talent Pool programmes for any career stage: engineering career is stable and predictable in the long term.

Youth influence in the aspect of overall management is carried out through the established and actively developing institute of Youth Councils. This tool is improving the professional life of young people.

Digital services basically form a value proposition for employees in terms of simplification of HR processes. All personnel processing and a significant share of employee training, engagement research, and other communication and document management are implemented in digital form.

Corporate engineering competitions, aimed to make an engaging environment for realizing creative experience in the professional sphere: ATOMSKILLS, ATOM-LAB, Person of the Year, Healthy lifestyle Ambassadors, Mission Talents, Technology leader, internal research projects, etc.

Career counselling shows a high degree of demand and satisfaction among Rosatom's engineers. The result of interaction between a career consultant and an engineer can be the scheme of desired career route, taking into account individual motives, profession and the current personnel situation at the enterprise and in the division, a career plan or an individual development plan that fills in the missing competencies.

Integration of talented engineers into the educational system in school/college/ university provides engineers an opportunity to become a professional teacher and prove yourself as a mentor for the younger generation.

Rosatom's corporate merch today is a format of native communications of the employer's brand, which allows to fit it seamlessly into the industry business agenda. The developed concepts are replicated at all of corporation subsidiaries, actively played out in social networks and at industry events. Today, the merch is a carrier of the idea of the company's desire for innovation and technological efficiency.

8. In addition to the tools described above for working directly with young engineers, we have implemented a number of additional adjustments in the internal HR management system:

- Motivation (financial and non-financial)

Building a competitive level of income for Rosatom's engineers related to the labour market metrics, quickly responding to changing market conditions. At the same time, the opportunities for increasing an employee's monthly income for mastering and developing competencies, developing, and implementing improvements, increasing personal and team productivity, and implementing unplanned or project tasks have been expanded. In terms of non-material motivation, a wide range of Rosatom and country level competitions for the best engineers.

- Success criteria for managers and teams

We provide multi-module training for managers at all levels on such topics as personal effectiveness, partnership, effective team, and innovation. The annual assessment of managers at all levels is supplemented by the "Human resource development" module and includes specific metrics of team success, the development of people entrusted to the manager, including young people.

- KPI system in the Human Resource Development and evaluation system

Building personalized annual goals for the manager, affecting among other things the level of income: focus on development, increasing labour productivity through the generation and implementation of proposals for process improvements, forming and adapting a team to new business environment conditions.

- Navigation systems for basic documentation based on key tags, visually clear instructions, one-page summary of voluminous documents

Intuitive navigation through the basic documentation has been built and working groups have been created – to describe regulatory documents in one-page format for initial review. Work instructions are drawn up, including using visualization tools. We are working on creating a chatbot for document navigation and expanding the use of artificial intelligence for document preparation and analysis.

9. The entire set of HR-efforts allows us to achieve results over a period of less than five years:

- Obtaining confirmed national status of Rosatom as one of the best employers;
- Growth in the rate of admission of young engineers more than doubled with youth share increase up to 22% backed by a 3-fold increase in the annual admission of young engineers with higher education and a similar annual growth rate in the promotion of youth to higher positions;
- Despite the entry of new young engineers and workers into the company every year, a high level of youth involvement of 85% is well maintained with an iterative growth of satisfaction for such factors as Training and development, Career opportunities, Reward and recognition, Work-life balance, which characterizes the high efficiency of the implemented activities for the category of young professionals;
- The turnover of young engineers is at an acceptable level and amounts to 13%, which indicates the need to keep improving the systemic integration of young engineers into the nuclear industry;
- It is important for us to integrate young people into the nuclear industry, relying on their best qualities and at the same time preserving the core historical values of the nuclear legacy.

C. Pesznyak
Associate Professor, Past President of the ENEN,
Budapest University of Technology and Economics, Hungary

C. Pesznyak spoke about the European Nuclear Education Network (ENEN) and its efforts to attract new talent and support lifelong learning in the nuclear field. ENEN, a nonprofit association with over 20 years of experience, coordinates EU-funded projects aimed at promoting nuclear education, research, and professional development. These initiatives include competitions for secondary school students, as well as programmes for bachelor's, master's, and PhD students, to encourage the next generation of nuclear professionals.

She highlighted the importance of ENEN2plus project, which involves 39 direct and 51 total partners. The project aims to enhance nuclear competencies through structured education and vocational training, with a focus on long-term sustainability. The project's key components include attracting new talent, facilitating policymaking, and supporting mobility schemes for students and early-career professionals.

She emphasized the critical role of public communication in increasing the acceptance of nuclear technologies, portraying nuclear energy as a clean, reliable, and sustainable solution. Academic education and research play an essential role in attracting young people to the nuclear field. ENEN's efforts also aim to ensure continuity in EU-funded projects, supporting both nuclear power and non-power applications, such as medical uses and radiation protection.

Furthermore, C. Pesznyak discussed the importance of vocational education and training, including mobility grants for students and professionals. These opportunities help participants gain international experience and cross-disciplinary knowledge, strengthening their commitment to the nuclear field.

Lastly, the speaker underscored the need for strategic activities and collaboration between universities, research institutes, and industry to maintain and develop a stable system for nuclear research, education, and training. She highlighted ENEN's efforts to foster relationships through PhD competitions and thesis awards, ensuring that young professionals stay engaged in nuclear science.

7.3.2. Session 13.1. University Education for Nuclear, Part 1

Session Chair: R. Garbil, EURATOM

IAEA Session Rapporteur: J. Roberts

The session focused on several key issues and topics related to nuclear education and workforce development across different countries. It addressed the development of international mobility schemes for nuclear education, including specialized master's programmes that cater to the growing global demand for skilled professionals. Discussions also covered the challenges faced by developing countries in establishing nuclear science programmes, including the need for significant human resource development to support nuclear energy and medical applications. International cooperation and partnerships, particularly with established nuclear organizations, were emphasized as crucial for building expertise. The role of mobility in enhancing nuclear talent was highlighted, as well as the importance of hands-on experience in diverse cultural and technical environments. Additionally, strategies for fostering the next generation of nuclear professionals, including collaborations between universities, research institutions, and industry, were explored.

Paper ID 407: The SARENA Programme: Education in Nuclear Engineering within a European Mobility Scheme – Feedback from Experience and Future Perspectives

Speaker: A. Abdelouas, France

SARENA, abbreviation for "SAfe and REliable Nuclear Applications", is a master's degree mobility scheme in the context of the Erasmus Mundus Mobility supported by European Education and Culture Executive Agency (EACEA). It relies on the master's degree programmes of its four consortium partners. This contribution presents an analysis of the feedback from experience from the first iteration concerning students' application and recruitment profiles as well as the professional insertion after obtaining the degree and satisfaction of the participating students.

Paper ID 177: Establishing Nuclear and Radiological Science Educational Programmes in Rwanda: Challenges and Opportunities

Speaker: C. Kwisanga, Rwanda

This paper delves into the main challenges of building nuclear and radiation programmes in the context of Sub-Saharan Africa, as well as the opportunities these programmes may represent nationally, regionally and at the continental level. Additionally, the paper outlines pathways for efficient and sustainable knowledge creation and retention within the country.

Paper ID 357: ROSATOM International Education Network

Speaker: E. Nagibina, Russian Federation

Today, universities and companies face a difficult task: to build new formats of interaction to ensure the socio-economic growth of their countries and enhance public acceptance of nuclear power. University networks and new interaction formats allow for the expansion of the educational and scientific infrastructure and ensure the transfer of the educational process and knowledge to partner countries to support development of national nuclear programmes.

Paper ID 333: Mobility in the Development of Nuclear Talents: Why and How

Speaker: L. Cizelj, Slovenia

This paper aims to display the current status of the education and training programme at CNSTN as well as the prospect of the Internet Reactor Laboratory. With the support of the IAEA's Internet Research Reactor (IRL) initiative, nuclear engineers and technicians from the Tunisian National Centre of Nuclear Sciences and Technologies can now take part in real-time reactor experiments at a reactor facility in France and, after that, in the Czech Republic.

Paper ID 46: Fostering Nuclear Professionals and Inspiring the Younger Generation by JAEA: Good Practices in Japan and Asian Countries

Speaker: Y. Nara, Japan

The JAEA offers a wide range of training programmes, from basic courses for working professionals to courses for expert development to foster skilled and qualified workforce. In addition, it actively engages in collaborative partnerships with universities and host interns so that students can acquire specific knowledge and practical skills and serve as the joint secretariat for the Nuclear Human Resource Development Network which involves industry, government, and academia in Japan. This approach is presented in the paper.

Paper ID 373: Development of a nuclear education strategy to meet human resource needs in Estonia

Speaker: A. Tkaczyk, Estonia

This paper will present the current initiatives in nuclear education and University of Tartu's efforts to cooperatively develop a nuclear education strategy and support the government in this complex matter. Although some training could be outsourced, it is appropriate to develop specific bachelor's and master's level programmes domestically in the long term.

7.3.3. Session 13.2. Education Networks

Session Chair: M. Hassan, Egypt

IAEA Session Rapporteur: H. Zhivitskaya

The session delved into global initiatives and networks advancing nuclear education and workforce development. It discussed European Commission-supported projects like ENEN2plus, OFFERR, and Tandem, addressing educational challenges in the nuclear sector. STAR-NET's role in fostering cooperation and specialized training among member universities is highlighted. The impact of ANENT in the Asia-Pacific region over two decades is examined, emphasizing collaborative learning and capacity-building. INSTA's establishment to enhance nuclear education in Asia-Pacific, and GNSSN's achievements in global nuclear safety and security, were also presented. Finally, AFRA-NEST's efforts in Africa to strengthen nuclear education and training through regional networks were discussed.

Paper ID 269: ENEN's Contribution to the Creation of a Competent Nuclear Workforce

Speaker: G.L. Pavel, ENEN

The paper explores the recent global resurgence in nuclear programmes, which has revitalized discussions about workforce development in the nuclear sector. This renewed interest is driven by proactive messages from the nuclear industry and supportive decision-makers, sparking increased interest among new generations in pursuing nuclear careers. Despite this positive trend, decades of decline in the nuclear industry and educational institutions pose challenges in meeting the workforce demands of new nuclear projects. The current human resource output is insufficient to address workforce attrition. The paper details the actions and outputs of several European Commission-supported projects, including ENEN2plus, OFFERR, Great-Pioneer, Tandem, ECC-SMART, and Go-Viking, aimed at bolstering nuclear education and training.

Paper ID 29: STAR-NET Continues to Build Competencies on Peaceful Application of Nuclear Technology

Speaker: A. Kosilov, STAR-NET

The paper considers STAR-NET, a Regional Network for Education and Training in Nuclear Technology, established to ensure sustainable human resources and promote safe nuclear technology. Since its inception in 2015, STAR-NET has fostered cooperation in education, professional training, research, and outreach. Key success factors include creating a common educational space, implementing distance learning platforms, and offering specialized training courses for member universities. This paper highlights STAR-NET's initiatives in enhancing nuclear knowledge and skills, crucial for the safe and sustainable utilization of nuclear technology worldwide.

Paper ID 404: ANENT's 20 Years of Learning and Sharing Nuclear Knowledge: A Historic Milestone

Speaker: S. Ariyanto, Indonesia

The paper examines the pivotal role of the Asian Network for Education in Nuclear Technology (ANENT) over the past two decades in advancing nuclear education and knowledge sharing in the Asia-Pacific region. ANENT's success is attributed to its comprehensive initiatives in education and training, promoting and preserving nuclear knowledge, and ensuring a steady supply of qualified human resources in the nuclear sector. Emphasizing collaborative learning, ANENT has facilitated the exchange of nuclear information, expertise, and best practices among member countries, enhancing nuclear capacity-building programmes. As ANENT celebrates its 20th anniversary, the paper highlights its journey marked by collaboration, knowledge sharing, and a commitment to advancing nuclear education in the region.

Paper ID 393: The International Nuclear Science and Technology Academy (INSTA): A New Initiative in Supporting Regional Nuclear Knowledge Management and Human Resource Development

Speaker: I. Abdul Rahman, Malaysia

The paper explores the role of the International Nuclear Science and Technology Academy (INSTA) in enhancing nuclear education and workforce development in the Asia-Pacific region. Recognizing the under-utilization of nuclear science and technology (NST) for economic growth due to a lack of qualified educators and workers, the IAEA's Technical Cooperation Asia and the Pacific section gathered experts in 2019 to devise solutions. The meeting highlighted the need for collaboration among universities, training centres, research organizations, and regulatory bodies. INSTA was established to address these gaps. The paper outlines INSTA's objectives, strategies, organizational structure, action plans, and potential collaborative avenues, emphasizing its complementary role alongside established educational networks linked with the IAEA.

Presentation: Introduction of the Global Nuclear Safety and Security Network (GNSSN): Achievements, Results and Perspectives

Speaker: K. Pavlova, IAEA

The presentation introduces the Global Nuclear Safety and Security Network (GNSSN), highlighting its achievements, results, and future perspectives. GNSSN serves as a collaborative platform promoting global nuclear safety and security through knowledge sharing, training, and technical cooperation. This presentation outlines key milestones, effectiveness in enhancing nuclear safety culture, and strategic directions for advancing international cooperation in safeguarding nuclear facilities and materials.

Paper ID 444: The African Network on Education in Nuclear Science and Technology – AFRA-NEST

Speaker: N. Hashim, Kenya

The paper examines the African Network on Education in Nuclear Science and Technology (AFRA-NEST), established during the 2007 AFRA Ministerial Conference in Aswan. AFRA-NEST, created to implement the AFRA strategy on Human Resource Development and Nuclear Knowledge Management, held its first General Assembly in 2013 in Arusha, Tanzania. The network, comprising national networks of education in nuclear science and technology (NEST), aims to enhance higher education, training, and research in Nuclear Science and Technology across Africa. AFRA-NEST also facilitates communication among member organizations and regional networks. Significant progress has been made in establishing national NESTs, with future steps focusing on developing national and regional e-learning platforms for educational and training applications.

7.3.4. Session 13.3. Attracting Young Generation

Session Chair: E. Pule, South Africa

IAEA Session Rapporteur: M. Ovanes

The session highlighted the importance of overcoming the lack of a skilled workforce and building diverse, gender-balanced, and inclusive teams in the nuclear industry. A survey by the Japan Atomic Industrial Forum, for example, revealed that only 28% of nuclear companies in Japan are meeting their recruitment needs, with concerns over the quality of hires. To attract young talent, several initiatives are being put in place across the nuclear sector, aiming to engage STEM students and bridge the gap between industry needs and workforce availability. Strategic workforce development plans include initiatives emphasising a robust education and training support, early engagement and diverse career pathways, gender diversity and increasing awareness to attract and retain skilled professionals in the industry.

Overall, the presenters conveyed the message that knowledge transfer and human resources development, which include educating, recruiting and developing skilled personnel, are essential for the sustainable use and progress of nuclear technology. The essential role of nuclear energy in climate change mitigation creates a compelling message for the youth to pursue careers in the nuclear sector. The success of attracting young talents to the nuclear sector hinges on our ability to present a compelling vision of a reliable future for nuclear energy.

Presentations of proSTEM Challenge 2024 Winner

The session also provided the opportunity to congratulate the IAEA proSTEM Challenge 2024 winners from UK (Sophie Osbourne) and Poland (Angelika M. Talaga and Dominika Gołąb) as well as to play videos presenting their projects: "Genius Journals" and "STEMTok – Widening STEM Outreach onto Social Media" respectfully.

Paper ID 382: Tabloo: Informing Pupils, Students and the Public about Nuclear Science and Technology

Speaker: M. Coeck, Belgium

The paper discusses the efforts of the Belgian Nuclear Research Centre (SCK CEN) in promoting nuclear science and technology through Tabloo, a visitor and meeting centre. Tabloo serves as a platform for informing pupils, students, and the public about nuclear science and technology. It houses several organizations and features a scientific exhibition and practical exercises for knowledge sharing. Visitors can participate in experiments, demonstrations, and interactive displays to learn about radioactivity, its applications, and the principles behind nuclear processes. The paper also highlights SCK CEN's interest in nuclear competence building and its efforts to develop a sustainable system that builds competences according to the needs of the institute and the country. The paper concludes by emphasizing the importance of science communication to inform and to promote STEM.

Paper ID 134: Growing the Next Generation Nuclear Talents in Bulgaria

Speaker: N. Velichkova, Bulgaria

Kozloduy Nuclear Power Plant (NPP) has implemented a well-structured workforce planning process and numerous initiatives to attract young talent. With over 140 employees retiring in the next decade, specific measures are in place to identify and preserve critical knowledge for the benefit of future nuclear experts. These measures aim to maintain key personnel competences, implement knowledge loss risk management and succession planning, maintain a high level of employee motivation, as well as rolling out outreach programmes to bring in the new generation of professionals. Within the past 5 years, 170 students from various universities took part in the scholarships programme at Kozloduy NPP that provides them with scholarship payment, labour contract upon graduation and granted paid internship.

Paper ID 242: Empowering Japan's Nuclear Future – Strategic Initiatives in Talent Acquisition for the Nuclear Industry

Speaker: T. Kita, Japan

In Japan, the nuclear industry encounters challenges in recruiting talent, especially due to the decreasing and aging population. To tackle these issues, strategic approaches are necessary. These include increasing interest in STEM subjects among junior and senior high school students, promoting understanding and acceptance of nuclear energy among both the younger generation and senior citizens, improving the teaching skills of junior and senior high school teachers through seminars, and providing young job seekers with opportunities to interact with companies and organizations in the nuclear industry. The Japan Atomic Industrial Forum, representing the nuclear industry in Japan, offers solutions and activities to address the challenges.

Paper ID 399: Impact of the Role of Nuclear Energy in Climate Topics on Attraction and Retention of Talents in Nuclear Sector

Speakers: J. Najder, European Nuclear Society, K. M. Madden, IAEA

The paper discusses the increasing visibility of nuclear-related organizations and initiatives that support youth participation in nuclear activities, particularly those related to climate change. Initiatives like Nuclear for Climate (N4C) have been successful in engaging youth and educating them about the role of nuclear energy in climate topics. The relation between engagement in climate topics and rising passion for nuclear energy was studied globally by IYNC's World Young Generation in Nuclear Thermometer (WYGNT) study. The study aims to provide valuable insights on demographic challenges and young professionals' perspectives, that can help employers in the nuclear sector attract and retain staff driven by their passion and sense of purpose.

Paper ID 411: General Manager People

Speaker: V. Barrins, Australia

ANSTO's strategic plan and engagement campaign aim to inspire the next generation towards nuclear and STEM careers, with a focus on increasing female representation. The campaign includes promoting diverse career paths, fostering women's participation, expanding educational support, strengthening networks with educational organizations, and broadening early career pathways. The plan comprises five key areas of focus: promoting the variety of career paths in the nuclear field; implementing inclusive policies and programmes to encourage women's participation in the nuclear sector; extending the support and training provided to the Australian tertiary and secondary education sector; establishing and reinforcing connections with associated educational organizations; and broadening ANSTO's early career pathways.

7.3.5. Session 14.1. University Education for Nuclear, Part 2

Session Chair: A. Aszodi, Hungary

IAEA Session Rapporteur: J. Roberts

The session focused on the development and implementation of nuclear technology management education programmes across several universities. Key topics included the establishment of specialized programmes to address the need for skilled leaders in the nuclear industry, with institutions from Bulgaria, Australia, Hungary, Russia, and the U.S. presenting their respective efforts. These programmes are tailored to meet International Nuclear Management Academy (INMA) standards, often developed in partnership with faculties of business and economics and aim to produce professionals capable of managing both large nuclear facilities and smaller modular reactors. Challenges such as workforce uncertainty, STEM interest decline, and the shortage of radiation specialists were discussed. A common theme was the need for global collaboration and knowledge sharing to ensure a sustainable nuclear workforce, with several universities considering English-language versions of their programmes to attract international students. Outcomes include the successful launch and growth of these programmes, with ongoing efforts to expand curriculum offerings and international cooperation.

Paper ID 92: Challenges in Development of Nuclear Technology Management Education Programme in Sofia University "St. Kliment Ohridski"

Speaker: P. Petkov, Bulgaria

The new challenges to nuclear industry in Bulgaria force Sofia University "St. Kliment Ohridski" constantly to improve the quality of higher education by developing of new educational programmes. Part of these efforts is the newly established programme, named "Nuclear Technology, Management and Innovations" (NTMI), developed jointly with the International Nuclear Management Academy (INMA). It is designed to educate students to become in future an interdisciplinary type of managers in nuclear industry, focusing on technology management for the nuclear sector including nuclear power, nuclear applications, and radiological technology. The current report gives an overview of the course curriculum as well as substantiates two possible future specializations: (1) large nuclear facilities (for power production) and (2) small modular reactors.

Paper ID 452: Nuclear Education and Knowledge Management Activities at the University of Adelaide, Australia

Speaker: T. Hooker, Australia

The Centre for Radiation Research, Education, and Innovation (CRREI) at the University of Adelaide is playing a pivotal role in addressing the growing need for expertise in nuclear radiation across various sectors in Australia. With the country's expanding involvement in nuclear activities, ranging from research and medicine to the development of nuclear submarines, there is a pressing demand for skilled professionals in this field. CRREI's initiatives aim to tackle two significant challenges facing Australia: attracting younger generations to STEM subjects and addressing the aging nuclear radiation workforce. By engaging with government, industry, and other educational institutions, CRREI is working to cultivate interest and expertise in nuclear radiation-related fields. This paper will discuss CRREI's current and future activities in providing education, training, and fostering collaboration among various stakeholders to enhance Australia's sovereign capability.

Paper ID 370: Lessons Learned from Nuclear Technology Management Education at BME and Future Development Plans

Speaker: M. Szieberth, Hungary

The paper will present lessons learned from the Nuclear Technology Management (NTM) post-graduate degree programme of the Budapest University of Technology and Economics (BME). A comprehensive evaluation of the programme based on a detailed survey of the past and present students, stakeholder consultation and the recommendations of the industrial advisory board will be presented and discussed.

Paper ID 40: Experience in Implementation of INMA Nuclear Technology Management Master's Programme at National Research Nuclear University MEPhI +Course for University Master's Level Programmes

Speaker: E. Kulikov, Russian Federation

The International Nuclear Management Academy master's programme in Nuclear Technology Management (NTM) at the NRNU MEPhI has been developed to improve safety, performance and economics of nuclear technologies by promoting and enabling the availability and accessibility of consistent high quality educational opportunities for nuclear sector managers and improving their management competencies. The programme is supported by the established links with the nuclear industry, including National Nuclear Cooperation ROSATOM and Nuclear Technology Education Consortium.

The NTM master's programme has been running at the NRNU MEPhI since 2016, during which time more than 35 graduates from Russia, Belarus, Kazakhstan, and Uzbekistan have received master's degree. The Faculty of Business Informatics and Integrated Systems Management and the Institute of Nuclear Physics and Technology are the key partners in the implementation of the programme. The paper discusses approaches in implementation of the NTM programme, key success factors and lessons learned.

Paper ID 443: The INMA-Endorsed Nuclear Technology Management Programme at the University of Idaho – Idaho Falls

Speaker: I. Charit, United States of America

The paper reviews the status of the INMA-endorsed Nuclear Technology Management Graduate Certificate Programme offered at the University of Idaho (UI). The courses need to be taken in conjunction with the Technology Management Master of Science or Nuclear Engineering Master of Science. INMA endorsed the programme in September 2023 after a visit by the INMA team to Idaho Falls during the summer of 2023. Since then, we are engaged in spreading the word about this coveted endorsement. The opportunity of obtaining an endorsement certificate from INMA has been conveyed to our current graduate students with the hope of attracting some students to be the first cohort following the endorsement attainment. The UI has a long-standing educational contract with the Idaho National Laboratory (INL) which is the leading nuclear energy national laboratory in the United States and is also located in Idaho Falls. We hope to develop new partnerships with other companies and governmental agencies.

Paper ID 214: Fostering the Next Generation of Nuclear Energy Sustainability Champions

Speaker: M. Gladyshev, IAEA

To facilitate the knowledge transfer, the IAEA's International Project on Innovative Nuclear Reactors and Fuel Cycles (INPRO) has developed a model curriculum for a master's degree course on strategic planning for sustainable nuclear energy development. The course's objective is to provide knowledge and practical skills in planning and modelling a nuclear energy system (NES) as scenarios evolve, and in using the INPRO methodology as a tool for performing sustainability assessment of NESs. The curriculum encompasses five competency groups:

- Energy planning and strategies for sustainable development;
- Planning for nuclear energy sustainability;
- Innovations in nuclear energy sector in meeting sustainable energy development challenges;
- Nuclear energy systems modelling and analysis;
- The methodology for assessing sustainability of nuclear energy systems (the INPRO Methodology).

Owing to the multidisciplinary nature of this topic, the number of higher education institutions (universities and centres of excellence) implementing this course in full may be limited. Therefore, this curriculum serves as a model and can be adopted and adapted in parts as needed to fit into respective educational programmes in interested Member States.

7.3.6. Session 14.2. Communities of Practice (CoP)

Session Chair: M. Roulleaux Dugage, France

IAEA Session Rapporteur: A. Ganesan

The session focused on the role of Communities of Practice (CoP) in fostering knowledge management within the nuclear sector. Presentations discussed the benefits of CoPs, such as enhanced knowledge sharing, cost savings, and innovation, with organizations like EDF reaping tangible results from over 200 CoPs. Other speakers highlighted how changes in governance and job area networks within nuclear companies can improve knowledge management and workforce development. The UK presented on tacit knowledge transfer processes, while the European Commission's Joint Research Centre emphasized the role of open access to research infrastructures in training a skilled workforce. The ENEN initiative NUCLEATION was showcased as a CoP aimed at strengthening European nuclear competences. The IAEA discussed the development of a TECDOC to support the effective formation and enhancement of CoPs across nuclear organizations.

Paper ID 167: NKM: Fostering a Sustainable Culture of Nuclear Knowledge Sharing through Standardised, Efficient, Active, and Impactful Nuclear Communities of Practice

Speaker: C. Chevereau, France

Over the past years, the EDF Group implemented an ambitious Nuclear Knowledge Management programme including several initiatives to promote a solid knowledge sharing culture as a key factor to nuclear safety and business performance. One of them fosters standardized, efficient and impactful Communities of Practice (CoP) i.e. groups of professionals who share knowledge around a specific topic. They may be as diverse as radiation-protection, operation under accidental conditions, hazards, transformers, I&C, welding, training, plant outage management etc., with 10 to 200 members each, located on various sites. EDF NKM programme supports the creation, improvement and maturity growth of these CoPs to make them capable of not only capitalizing, sharing and disseminating nuclear knowledge and experience feedback, but also innovating and producing value. For this purpose, EDF implemented a set of standards, best practices for CoPs, as well as training in terms of organization, sponsoring, animation, assessment and performance-oriented annual objectives as well as a meta-CoP liaising all the nuclear CoP leaders. With now 200+ nuclear CoPs, some being over 15 years old, EDF is now reaping tangible benefits from them, in terms of increased access to knowledge for its workforce, including new recruits, savings, innovations and cost avoidance.

Paper ID 387: How EDF Nuclear Generation Division Governance Creates Favourable Conditions to Develop and Promote Knowledge and Skills Sharing Amongst the Nuclear Fleet

Speaker: C. Regnaud, France

The paper illustrates how the recent changes in EDF nuclear corporate's governance impact knowledge management within the fleet. It provides insights on the benefits gained from a new KM project launched in year 2020. It highlights how the new governance based on core job areas (occupations) such as operation, maintenance, supply chain, etc. influence KM objectives.

Paper ID 324: Enabling Tacit Knowledge Transfer through Communities of Practices and Retention of Critical Knowledge in NWS

Speaker: H. Malik, United Kingdom

The paper discusses the approach to Knowledge Management in NWS, NDA UK, focusing primarily on people and processes. It also talks about the delivery of tacit knowledge capture from existing and departing staff through Communities of Practices and Knowledge Retention processes. The paper provides examples of supporting policies, processes, methods, and communication being introduced in both fields. The paper discusses the integration with Organisational Learning in NWS to give added value to holistic knowledge transfers and learning with KM.

Paper ID 384: A European perspective – The Nuclear Research Infrastructures Open Access Scheme of the Joint Research Centre (JRC) at the European Commission – Contributions to Education, Training, Mobility and Scientific Excellence

Speaker: A. Siebert, European Commission

Regardless of a country's position on nuclear in their energy mix, there will be a need for competent and well-trained people in the nuclear field addressing societal challenges such as energy, health or environmental applications or other non-power nuclear applications. To meet these demands, it is essential to preserve and develop expertise in various nuclear fields. The Joint Research Centre (JRC), as a Directorate-General of the European Commission contributes to this task by providing training, education and access to its nuclear laboratories. The paper describes how the JRC plays an important role in training and developing a competent workforce in the nuclear field specifically through its Open Access to Research Infrastructures (RI) programme. The Open Access RIs comprise state of the art facilities including laboratories for preparation, characterization and physical studies of actinide bearing materials, hot cells, neutron sources, metrology laboratories, and mechanical and corrosion properties test laboratories located on three different JRC sites. Access modalities, success stories and lessons-learned are discussed.

Paper ID 146: NUCLEATION – a Community of Practice on Vocational Training in Nuclear Domains

Speaker: C. Schönfelder, ENEN

The paper presents the embedding of NUCLEATION in the ENEN2plus project, introduces the concept of a community of practice, and describes the characteristics of NUCLEATION. As this community can significantly contribute to build and enhance European nuclear competences, people from the target groups are invited to join NUCLEATION.

Presentation: Introduction to IAEA TECDOC on CoP

Speaker: A. Ganesan, IAEA

This presentation briefly explains the objectives and scope of the IAEA initiative to produce a TECDOC is aimed to help MS nuclear organizations.

- Develop a common understanding and theoretical framework for describing Communities of Practice (CoP) and its benefits in the nuclear context;
- Provide practical guidance on strategy formation and action planning to effectively support, promote and enhance the performance of CoP in Nuclear organizations;
- Identify examples of good practice and lessons learned from applying CoP principles;
- Describe the diversity of CoP arrangements in member states.

This presentation explains the structure and contents of the document.

7.3.7. Session 14.3. Outreach Education Programmes

Session Chair: I. Abdul Rahman

IAEA Session Rapporteur: H. Zhivitskaya

This session explored diverse aspects of nuclear education and workforce development, highlighting initiatives such as the CEATEN programme in Argentina and the SCK CEN Academy's efforts in Belgium. It discussed motivational factors influencing STEM students' interest in nuclear courses, analysed through Structural Equation Modelling. The session also covered Rosatom's comprehensive approach to nurturing scientific talent. The session inaugurated the International Nuclear Science Olympiad (INSO) aimed at fostering global nuclear education. Additionally, the session examines Malaysia's strategic integration of national curricula to meet industry demands, emphasizing collaboration with the IAEA. These efforts underscored the session's focus on enhancing educational foundations and preparing a sustainable nuclear workforce.

Paper ID 401: CEATEN: the Postgraduate Course that Brings Nuclear Knowledge to Professionals from all Branches of Science and Engineering

Speaker: F. Cantargi, Argentina

The paper discusses the Specialization Course in Technological Applications of Nuclear Energy (CEATEN), a one-year postgraduate programme offered through a partnership between Argentina's National Atomic Energy Commission (CNEA), National University of Cuyo (UNCuyo), and Buenos Aires University (UBA). Running since 1995, CEATEN selects students based on academic merit and provides fellowships from CNEA and partner companies like INVAP and Nucleoeléctrica Argentina (NA-SA). The course comprises expert-led modules at various CNEA locations in Buenos Aires and Bariloche. Students from diverse backgrounds, including Argentina, other Latin American countries, and Europe, participate. Graduates, exceeding 250, have often advanced to middle and senior management roles in their organizations.

Paper ID 281: Nuclear Competence Building by the Belgian SCK CEN Academy

Speaker: M. Coeck, Belgium

The paper highlights the Belgian Nuclear Research Centre (SCK CEN)'s commitment to nuclear competence building, crucial for maintaining a skilled workforce in various sectors. Leveraging its extensive R&D experience, innovative projects, and unique facilities, SCK CEN plays a pivotal role in nuclear education and training. The SCK CEN Academy engages with students, teachers, and the public to inform them about radioactivity and related career opportunities, supervises students and junior researchers, contributes to academic courses, and offers professional training. Additionally, it supports education policy and fosters collaborations and projects. The Academy also integrates social sciences and humanities into its initiatives. This presentation outlines SCK CEN Academy's vision and activities to ensure future nuclear sector competencies.

Paper ID 310: Examining Key Factors Shaping Youth's Intentions in Pursuing Nuclear-Related Courses: Attitude Analysis for Workforce Sustainability

Speaker: Z. J. Belmonte, Philippines

The paper investigates the motivations and intentions of young individuals in STEM tracks to pursue nuclear-related courses, crucial for fostering a skilled workforce in the nuclear industry. Using Structural Equation Modelling and theories like Self-Determination Theory and Theory of Planned Behaviour, the study identifies key factors influencing these intentions, such as attitudes towards nuclear-STEM, subjective norms, and perceptions of autonomy and competence. Findings suggest tailored strategies are needed to enhance education and outreach efforts, addressing generational disparities in perceptions of nuclear energy. This research provides valuable insights for nuclear workforce management, advocating for improved education strategies to cultivate a diverse and knowledgeable workforce, particularly in developing nations.

Paper ID 319: System of Attracting and Development Scientific Personnel for Advanced Technology

Speaker: E. Rakhmankina, Russian Federation

The paper explores Rosatom's multifaceted approach to nurturing scientific talent across various sectors, including nuclear medicine, materials science, and digital solutions alongside nuclear power. Beginning with early education initiatives in kindergartens and schools, Rosatom employs a comprehensive training system that includes internships and Science Schools for promising students mentored by experts. This strategy aims to foster scientific discovery and leadership among graduates, with 36% already achieving significant scientific breakthroughs. Key methods include early career orientation, university partnerships, and facilitating smooth transitions into the workforce, supported by scientist mobility and maintaining age-balanced teams. This approach ensures the continuity of critical competencies and enhances collaboration across generations within scientific institutes.

Paper ID 272: Inaugural International Nuclear Science Olympiad: Fostering Excellence in Youth Education

Speaker: M. Al'Azri, Oman

The paper introduces the First International Nuclear Science Olympiad (INSO), initiated under the auspices of the IAEA TC Project RAS0091. Planned for 2024, INSO represents a pioneering effort to promote excellence in nuclear science education among youth worldwide. Developed through collaboration among experts, INSO aims to inspire and challenge young minds through a structured curriculum and competitive platform. More than a mere contest, INSO fosters international camaraderie and collaboration, preparing participants for careers in nuclear science and engineering. With a focus on innovation and curiosity, INSO aims to enhance the educational landscape by providing a stimulating environment for young enthusiasts passionate about nuclear science. This abstract highlights INSO's role in shaping the future generation of nuclear scientists and engineers.

Paper ID 173: Building a Sustainable Nuclear Workforce: Strengthening the Foundation through National Curricula

Speaker: H. Adnan, Malaysia

The paper examines the pivotal role of national curricula in shaping a sustainable nuclear workforce amidst the evolving demands of the industry. Focusing on Malaysia, it explores strategic approaches to enhance educational foundations in secondary schools, aligning with industry needs for innovation and sustainability. Collaborative efforts between the Ministry of Education, Malaysian Nuclear Agency, and support from the IAEA's technical cooperation projects (RAS0065, RAS0079, RAS0091) are highlighted. The discussion emphasizes tailored teaching methods, educational materials, practical exercises, and co-curricular activities aimed at fostering student engagement with nuclear concepts. This holistic approach aims to prepare a well-prepared workforce capable of meeting the future challenges and opportunities in the nuclear sector.

7.3. MODERATED PANEL DISCUSSION ON GLOBAL ALLIANCES BETWEEN EDUCATION AND INDUSTRY

Panel discussion was moderated by J. Sowagi, Director, Information Exchange, CANDU Owners Group Inc. (COG), Canada.

Panellist:

M. Rolleaux Dugage, Consultant, MOPSOS, France

H. Malik, Knowledge Lead Partner, Information Security and Governance, Nuclear Waste Services, UK

N. Bonilla, Deputy Head of Division for Education Programmes, OECD Nuclear Energy Agency

Y. Wang, Vice President, Nuclear Industry College of the China National Nuclear Corporation, China

L. Cizelj, Head of Reactor Engineering Division, Researcher, Professor of Nuclear Engineering, Jožef Stefan Institute, University of Ljubljana, Slovenia

Panel discussion focused on exploring the critical role of collaboration in addressing workforce needs in the nuclear sector. The session highlighted how partnerships between educational institutions and the nuclear industry can help close the gap between academic training and industry demands. The discussion covered strategies for aligning education with industry needs, attracting young talent to STEM fields, and overcoming workforce shortages. It also addressed the role of government and other stakeholders in facilitating these alliances.

The following key points were noted during the panel discussion.

- Benefits of Alliances Between Education and Industry. The panel emphasized that partnerships between the nuclear industry and educational institutions are essential for aligning curricula with workforce demands. These alliances enable the industry to communicate its human resource needs, both short and long-term, directly to educational institutions. In turn, academia can adjust its offerings to match the skills and knowledge that the industry requires. Mentorship programmes, site visits, and other engagement activities were also highlighted as effective ways for students to connect with the industry and develop clear career paths;
- Addressing Workforce Shortages. The panel discussed how collaboration between industry experts and universities can bridge the gap between education and industry needs. By allowing industry professionals to lecture at universities and engage with students, these partnerships provide a way to align academic training with the practical needs of the nuclear sector. Additionally, this collaboration offers companies a chance to identify and recruit talented students through internships and other programmes, helping to address workforce shortages in critical areas;
- Challenges in Forming Global Alliances. The panel identified several challenges in forming and maintaining global alliances between education and industry. One major challenge is the differing objectives and timeframes of academic institutions, regulatory bodies, and industry. Cooperation between these groups often requires external pressure to ensure effective communication and collaboration. Another challenge is ensuring that academia's long-term, research-driven focus is aligned with the industry's more immediate, operational needs;
- Aligning Educational Curricula with Industry Needs. The discussion highlighted the importance of educational institutions regularly updating their curricula to reflect the evolving needs of the

nuclear industry. Collaboration with industry professionals can provide insights into current and future workforce requirements, ensuring that students receive relevant training. Programmes that combine theoretical education with practical experience, such as internships and mentorships, were cited as effective ways to ensure that students are well-prepared for careers in the nuclear sector;

- Successful Collaboration Programmes. Examples of successful collaborations between academia and industry were shared, particularly those that create knowledge hubs or innovation centres where students, educators, and industry professionals can exchange information and ideas. The pandemic accelerated the use of digital learning tools, making it easier for education and industry to collaborate remotely. These programmes allow for flexible, real-time collaboration, enabling a more dynamic learning and training environment;

- Attracting Young Talent to the Nuclear Industry. The panel discussed the need to attract more young people, especially women, to STEM fields and the nuclear industry. Mentorship programmes targeting high school students, particularly girls, were highlighted as a successful strategy. These programmes allow students to engage with industry professionals, hear about their experiences, and address concerns about working in a male-dominated field. The panel also pointed out that young professionals are increasingly drawn to innovative projects, such as Small Modular Reactors (SMRs), that align with broader societal goals like climate change mitigation;

- The Role of the State in Education-Industry Alliances. Governments play a key role in facilitating alliances between the nuclear industry and educational institutions. In some countries, government initiatives bring together universities, job agencies, and the nuclear industry to align educational programmes with industry needs. These initiatives help students transition from education to employment by offering scholarships, internships, and guaranteed job placements, ensuring that graduates are ready to meet the demands of the nuclear workforce;

- Challenges in Filling Crucial Roles. Certain fields within the nuclear industry, such as electrical engineering and Instrumentation & Control (I&C), are experiencing significant workforce shortages. The panel stressed the need for better communication between educational institutions and industry to raise awareness of these crucial but lesser-known roles. Public awareness campaigns that highlight the connection between nuclear energy and important societal issues, such as climate change, were suggested as a way to attract talent to these vital disciplines.

The panel concluded that global alliances between education and industry are critical to addressing the workforce challenges facing the nuclear sector. These partnerships allow for better alignment of curricula with industry needs, provide clear career pathways for students, and foster collaboration that benefits both academia and the nuclear industry. Governments and other stakeholders also have an important role in supporting these alliances. Looking forward, these partnerships will be essential in ensuring that the nuclear industry remains competitive, innovative, and able to meet the growing demand for clean energy and skilled professionals, especially as the industry moves toward the goal of net-zero emissions by 2050.

8. SUSTAINABILITY: LOOKING FORWARD, RESILIENCE AND PROGRESS

8.1. KEYNOTE SESSION

As prepared for delivery.

J. Gadano
Vice President, Nucleoeléctrica Argentina S.A., Argentina

Good morning, Chair and dear colleagues.

My name is Julián Gadano, I am from Argentina. I have worked in the nuclear sector for 15 years, and I am currently the Vice President of Nucleoeléctrica Argentina, which is the operator of the 3 nuclear plants we have in our country. I intend to talk to you today about what, faced with the scenario that is coming to us, can be the difference between success or failure. I am referring to the very probable nuclear expansion, hand in hand with the growth of new generation reactors (which, let me insist, we have to understand more as a new business model than as a technological innovation).

Let me begin with a relatively broad and general definition of our reality today. In today's world, humanity faces unprecedented challenges, placing us at a pivotal historical moment. Nuclear energy plays a crucial role in addressing these challenges. Climate change, the instability of energy markets, the unprecedented war in Europe and other global issues demand safe, clean, reliable, and cost-effective electricity generation. These factors strongly motivate nuclear power plant operators not only to extend the operating life of their facilities but also to expand nuclear activities with strong capabilities.

Moreover, the Sustainable Development Goals highlight nuclear energy and its applications as strategic assets, tools, and resources to address global challenges. Anyway... we are used to hearing that the time for renaissance, for growth, has finally arrived... The new thing and the good thing are that, in my opinion, now it is for real. We observe a renewed interest from IAEA member states in nuclear power to achieve net-zero emissions and ensure energy security in the face of a world in which conflicts are growing.

The IAEA projections indicate that the share of nuclear power could represent up to 14% of the electricity mix by 2050, a substantial increase from the current figure of 9.8%. And in my opinion, the growth prospects for nuclear energy due basically to the SMR's expansion, in markets such as Eastern Europe and Africa in the next 20 years, indicate that the agency's forecast is extremely conservative.

Among the 50 countries that have expressed interest in introducing nuclear power, 24 are in a pre-decision phase and engaged in planning activities, while the remaining 26 countries are actively pursuing the introduction of nuclear power. By 2035, the number of countries operating NPPs may increase by about 30%, with an additional 10–12 countries operating NPPs compared to the current 32. This growing interest in nuclear power needs the development of adequate nuclear infrastructure.

Collaborative efforts among nations, facilitated by international organizations, will be essential in sharing knowledge, expertise, and best practices to ensure the safe and efficient use of nuclear energy worldwide.

Reflecting on the evolution of nuclear power over the decades, we recognize that the first commercial nuclear power plants (NPPs) began operation in the 1950s, heralding a new era of energy generation. The 1960s and 1970s marked a significant expansion of nuclear activities, leading to the creation of nuclear clusters built on pillars such as industrial infrastructure, science and technology agencies, state promotion, universities, and private sector support.

The world has been able to operate nuclear facilities efficiently, safely, and securely, fostering a nuclear cultural background that we need to protect and continuously promote.

However, after these golden decades, the nuclear industry faced more complex times. There was a significant decrease in new project expansions, challenges in recruiting new professionals, the need for innovative capacity building and training schemes, and the retirement of experienced professionals. Today, we find ourselves in a situation of ambiguity: we are experiencing a period of "nuclear growing" with a greater demand for trained human resources, while simultaneously witnessing the retirement or emigration of senior experts, leading to a loss of valuable knowledge and experience.

This scenario offers both opportunities and significant challenges:

(1) Ambitious nuclear programmes that require highly trained resources to execute and maintain them in the long term;
(2) Retirement of experienced professionals;
(3) A huge generational gap, with young graduates needing to collaborate with those nearing retirement;
(4) Migration of human resources to other sectors such as oil and gas, or renewable energy.

Today, it is crucial to ask ourselves if we are prepared to attract new talents or at least to maintain our technical staff within the nuclear sector. Do we have sufficient incentives to train, motivate, and retain them? Can we promote and establish clear mechanisms at both national and international levels to encourage good professionals to work in nuclear?

The development of policies aimed at engaging youth in our activities is a crucial pillar for capacity building in the nuclear field. The generational gap is a shared reality among all represented countries, and it can only be bridged through strategic policy planning and the adoption of effective systems for knowledge transfer and management.

Consider this premise: every nuclear project – even projects based on small reactors – entails a medium to long-term commitment involving a diverse range of stakeholders. Effective management tools are essential to ensure the sustainability of these projects. Engaging and fostering the participation of new generations in nuclear activities is an unprecedented challenge today. Addressing this challenge requires exchanging best practices and promoting strategic collaborations among public, private, academic, and industrial sectors to develop sustainable technological projects with a strong emphasis on youth involvement.

In recent years, Argentina has fostered a significant awareness of the role of knowledge management, particularly since the re-launch of the Nuclear Plan in 2006. This initiative faced formidable challenges, including restarting construction on the Atucha II Nuclear Power Plant, and the life extension project of the Embalse nuclear power plant, an LTO for the next 30 years at full power (a project that I had the honour of leading). This project was unprecedented, given its scope and the innovative nature that we had to address.

They both also aimed to demonstrate expertise and know-how in critical nuclear fuel cycle activities such as nuclear fuel fabrication. Furthermore, efforts were made to repatriate Argentine scientists and highly specialized professionals, encouraging their return to a sector characterized – in the immediate past at the starting of the projects – by instability and lack of incentives. It simply implied nurturing a robust nuclear knowledge community after years of stagnation.

Beyond the nuances of each stakeholder within the nuclear ecosystem, the pursuit of a cohesive organizational culture has proven pivotal. This culture has emphasized the importance of Introductory

Concepts to Nuclear Knowledge Management, Nuclear Information Management, and the role of individuals and knowledge mapping in effective management practices.

Policies, strategies, and organizational cultures are essential for our approach to promoting dialogue and collaboration in Argentina's nuclear sector, which provides over 10,000 direct jobs. This approach encouraged individuals to take responsibility in technical, administrative, and academic roles. We understand that organizational growth depends on the knowledge and interactions of our people. Therefore, we prioritize fostering meaningful human connections that drive ongoing improvement and innovation.

However, once this scenario of growth in our capabilities in terms of knowledge and human resources has ended, the problems have reappeared in our country (which I consider to be just a sample of what is happening in the world). The construction model of large nuclear reactors, requiring a lot of CapEx, faces many difficulties currently. The good news is that innovation gained ground in the industry, and a new business model appeared: the acronym SMR means much more than small, modular reactors. It means the birth of a new business model based on reactors that will be built industrially based, with mainly private financing and at times and costs that promise to be competitive. The number of projects, in different stages of development allows us to assume that the change is sustainable and robust.

This scenario of a new nuclear industry that is growing vigorously on the one hand, and professionals seeking new horizons outside the nuclear sector on the other, implies a paradox that requires our attention:

On the one hand, we face (even some countries that have had projects underway) a shrinking workforce of professionals, who are looking for alternatives in other fields. And on the other hand, more than 25 companies around the world (most of them in the US and Canada) developing new reactor models. Some of these companies have been in the market for many years, but the vast majority are completely new. They didn't exist 15 years ago. On the one side, there is an enormous demand for qualified human resources, and on the other, there are professionals and technicians with nuclear skills looking for new horizons for their professional progress. This is real, it is happening today. It is our duty to connect both dots. How to do it?

First of all, knowledge should be preserved. I take this opportunity to recognize the efforts of those who organize this event, and also of those who from national institutions sometimes work without all the necessary resources to preserve and build knowledge. It is important to have solvent and funded institutions and organizations that preserve nuclear knowledge.

Secondly, we have to strengthen knowledge networks. The relationship between universities, industry, the scientific sector and the government is essential. I remember in my student days reading one of the founding fathers of our nuclear sector, Jorge Sabato, about the imperative need for this cooperation. Today I feel it as a very palpable reality. There are very virtuous examples all over the world. In our American continent, the clearest cases of how this cooperation produces knowledge are the United States (very clearly) and Canada. But in Latin America we also have successful cases in Brazil, Argentina and other countries.

Third, we have to understand that the nuclear industry is changing radically. It has become a much more global industry, and no project is resolved within the borders of one country. No country produces everything and no country buys everything. We have to lose our fear of the globalization of knowledge. "Protecting" the knowledge linked to the peaceful uses of nuclear energy shrinks us as a sector and impoverishes us as a society. I'm not talking about giving away anything for free: knowledge has value, related to the investment made to build it and the need to continue investing. But that is one thing, and acting endogenously is quite another. There are people who know how to do things, and there are companies that need that knowledge. An effort is required on the part of governments, universities and civil society organizations to reduce the gap between both worlds.

To conclude, it seems opportune to propose a positive scenario and focus on what we can do today: leveraging existing mechanisms and tools to enhance certainty and build capabilities, while bolstering infrastructure and elevating this issue on the international agenda of key stakeholders in our field.

Adopting a situated approach is crucial for robust risk analysis concerning knowledge retention, particularly coming from countries like Argentina with over seven decades of a distinctive development in the Latin America and Caribbean. Working closely with government officials and decision makers is crucial because effective actions depend on institutional policies that consider both past experiences and future plans.

In this context, strategic initiatives should incorporate:

- Extensive international cooperation: recognizing disparities in nuclear development among newcomer, operating, expanding countries, and those with established educational institutions and longstanding programmes. Enhancing exchanges, contacts, and joint participation in Agency's projects is paramount. Moreover, bilateral collaboration can capitalize on excellent educational institutions in represented countries to forge strategic training and educational partnerships;
- Establishment of 'Educational Networks': creating formal and informal frameworks to strengthen regional and interregional collaboration. This includes optimizing resource sharing, leveraging capabilities, fostering synergies in sustainable nuclear curricula, promoting student and faculty exchanges, and facilitating the dissemination of nuclear knowledge. These efforts aim to raise public awareness and attract and retain talented youth;
- Promotion of youth development policies by the private sector: initiatives such as science, innovation, and engineering competitions supported by industry play a pivotal role in keeping young talents engaged and fostering technological spin-offs;
- Integration of the topic into international forums and initiatives: while the Agency provides essential guidance and expertise, there's a burgeoning opportunity within various forums, agencies, and multilateral initiatives to push this agenda by bridging public, academic, and industrial sectors;
- Accepting international cooperation and exchange as an essential part of the Game.

It's today, it's now. The world is saying "nuclear energy, we need you" but hey! Change the mindset.

G. Bikkulova
Deputy Director General, Director of the International Initiatives and Partnerships Division,
Rosatom Corporate Academy, Russian Federation

The speaker highlighted the innovative training programmes developed by the Rosatom Corporate Academy, which are designed to build the necessary skills for the nuclear workforce of the future. These programmes incorporate skills intelligence, leadership development, and focus heavily on sustainability, aligning with Rosatom's long-term strategy. By analysing data and reports, Rosatom forecasts the skills that will be required and prepares its workforce through continuous learning and capacity-building initiatives.

G. Bikkulova stressed the importance of sustainability and resilience as core concepts embedded in Rosatom's HRD approach. With 370,000 employees across 460 enterprises, Rosatom tailors its HR strategies to meet diverse needs, while also ensuring the well-being of employees' families, particularly in the 31 towns where Rosatom plays a central role. This human-centred strategy goes beyond just employment, with a focus on social infrastructure and the holistic development of employees through education, well-being programmes, and leadership training.

One of the key themes of her presentation was the integration of stakeholder relationships and community engagement in HRD strategies. G. Bikkulova emphasized the evolving landscape of stakeholder demands, particularly the growing need for transparency, inclusivity, and sustainability in business practices. For Rosatom, this meant engaging with local communities, universities, and other organizations to create educational opportunities and economic growth. Their comprehensive training system spans from schools to universities, aiming to develop a highly skilled workforce that can meet the demands of the nuclear sector, both in Russia and in the international markets where Rosatom operates.

Rosatom's Human-Centric Corporate Philosophy, introduced in 2018, was another cornerstone of their HRD and NKM approach. This philosophy prioritizes the well-being of individuals and teams, ensuring that every business operation is aligned with fostering personal and professional growth. G. Bikkulova described how Rosatom's employee development programmes include e-learning platforms, mental health monitoring, and well-being initiatives, ensuring a resilient and adaptable workforce. These efforts are supplemented by mentorship programmes, where retired employees play an active role in guiding younger generations.

Rosatom has extended its human-centric approach to its educational and training partnerships. Collaborations with 20 Russian universities, including MEPhI, ensure that future nuclear professionals are equipped with the knowledge and skills necessary for the evolving nuclear landscape. Additionally, the Human-Centric Alliance, which G. Bikkulova introduced, is a platform where businesses from ten countries collaborate to develop advanced human-centred solutions, ensuring that knowledge-sharing and capacity-building extend beyond Rosatom to the broader nuclear and energy sectors.

G. Bikkulova concluded by addressing the critical role of sustainability in Rosatom's HRD and NKM efforts. Recognizing that sustainability education often focuses on challenges rather than solutions, Rosatom integrated practical sustainability thinking into its training programmes, helping employees and communities adopt a mindset of growth and adaptation. Through initiatives like weekly webinars, online courses, and sustainability education for local communities, Rosatom has fostered a culture of innovation and resilience, crucial for the future of the nuclear industry.

K. Mrabit

President of PAM Parliamentary Group, House of Councillors, Morocco

Human Resource Development for Nuclear and Radiological Safety and Security:

A National Strategy for Education and Training

In the keynote speech, Member of the Moroccan Parliament, and former Director General of the Moroccan Agency for Nuclear and Radiological Safety and Security (the Regulatory Body AMSSNuR) reflected on his extensive experience and highlighted the establishment of AMSSNuR. He discussed Morocco's strong commitment to renewable energy, with ambitious targets of over 50 % of its installed capacity for 2030 and noted that nuclear energy is being considered as an alternative option beyond that date. He emphasized the critical importance of education and training in nuclear and radiation safety to establish, implemented and continuously improve nuclear and radiological applications.

The keynote speaker outlined Morocco's national strategy for education, training, and knowledge management in the areas of nuclear safety, security, and safeguards (the 3S). He described the evolutionary structural changes in Morocco's legal and regulatory framework, particularly the creation of an independent regulatory body in 2016 that reports directly to the Head of Government (Prime Minister). This shift allowed Morocco to comply with relevant International Instruments and better align with international safety standards and security guidance, in close cooperation with the IAEA and the European Commission as well as other IAEA donor countries.

The strategy on Education and Training was established in collaboration with over 30 national partners, including universities, technical support organizations, and other stakeholders involved in nuclear and radiological applications. The IAEA, inter alia, played a significant role in supporting these efforts, providing expertise and guidance on safety standards and security guidance, which helped build a robust system for knowledge management and capacity building in Morocco.

Additionally, the development of capacity-building centres, particularly in emergency planning and response and nuclear security, was emphasized. These centres, established in partnership with the IAEA, were designed to support national activities and contribute to regional efforts, especially in Africa. The importance of international cooperation was highlighted, with partnerships formed through the establishment and implementation of memoranda of understanding with numerous countries and organizations.

While establishing such a broad collaborative system, some challenges were faced, especially in coordinating with government ministries, industry, and NGOs. But the final outcome was a robust and sustainable legal and regulatory framework integrating education, training, and knowledge management at an early stage.

In conclusion, the speaker emphasized the need for sustainability in nuclear and radiological education and training efforts, as well as continuous improvement process. The national strategy on E&T was established with and implemented by relevant government institutions, industry, Technical Support Organization, Operators and other relevant stakeholders. With the establishment and implementation of its Integrated Management System (IMS), the country ensured safe and secured nuclear and radiological applications. All the Moroccan capabilities in nuclear and radiological safety and security are made available to the other IAEA African Member States.

8.2. MODERATED PANEL DISCUSSION ON SUSTAINABLE STAKEHOLDER ENGAGEMENT IN NUCLEAR PROGRAMMES

Panel discussion was moderated by I. Chatzis, IAEA.

Panellists:

H. Adnan, Director of Information Management Division, Malaysian Nuclear Agency, Malaysia

W. Ndubai, Director for Strategy and Planning, Nuclear Power and Energy Agency (NuPEAA), Kenya

A. Aszodi, Professor and Dean, University of Technology and Economics, Hungary

V. Kulmala, Communication Expert, TVO, Finland

The discussion underscored the complexity of stakeholder engagement in nuclear power projects, emphasizing the need for transparency, tailored communication strategies, and the proactive management of public concerns and misinformation. The experiences shared by the panellists highlighted different approaches to achieving sustainable stakeholder engagement, with an emphasis on education, infrastructure development, and the strategic use of media. During the session panellists also addressed questions from audience and further elaborated on their strategies and experiences, providing valuable insights for other countries embarking on or managing nuclear power projects.

The following key points were noted during the panel discussion:

- Importance of Education and Continuous Engagement. The panellists stressed the importance of ongoing education and capacity building, even in countries like Malaysia, where nuclear power is not yet part of the national energy policy. Ensuring that the workforce and the public are prepared and informed is crucial for maintaining readiness should the government decide to pursue nuclear power in the future;
- Customized Stakeholder Engagement Strategies. The discussion highlighted the need for tailored engagement strategies that address the specific concerns of different stakeholder groups. For example, in Kenya, misinformation from anti-nuclear groups has required a re-evaluation of engagement tactics, particularly in local communities where fears about environmental and health impacts have been fuelled by incorrect information;
- Responding to Non-Safety Related Concerns. A. Aszodi from Hungary provided an insightful example where the primary concern from a local community near the Paks 2 NPP was not about nuclear safety but about the need for a bridge. This example underscores the importance of understanding and addressing the broader social and infrastructure needs of communities affected by nuclear power projects;
- Public Transparency and Access to Information. The panellists emphasized the importance of transparency in managing public perception and building trust. This includes making relevant environmental and safety data publicly available, as seen in Hungary's approach, where all environmental impact assessment documents were published online in multiple languages;
- Dealing with Anti-Nuclear Activism. The panel acknowledged the challenges posed by well-funded and organized anti-nuclear groups. Panellists discussed the importance of proactive engagement and the need to counter misinformation with accurate and consistent communication. The panel also noted that in some cases, the objective of engaging with activists is not to change their views but to inform the broader public and ensure that the legal and procedural integrity of public hearings is maintained;

- Leveraging Social Media and Influencers. The role of social media and local influencers in shaping public opinion was discussed. Panellists shared their experience in engagement with influencers to reach wider audiences and improve public understanding of nuclear power;
- International and Cross-Border Engagement. The panellists also touched on the importance of engaging with neighbouring countries, especially in cases where nuclear projects might have cross-border environmental impacts. For instance, Hungary involved 30 countries in the environmental licensing process for the Paks 2 project, ensuring transparency and preventing criticism by addressing concerns from the international community.

In conclusion, the panel underscored the complexity of stakeholder engagement in nuclear programmes, highlighting the need for transparent communication, tailored engagement strategies, and proactive management of misinformation. The experiences shared by the panellists demonstrated the importance of understanding the diverse concerns of stakeholders, addressing them effectively, and ensuring that the benefits of nuclear power are clearly communicated to the public.

8.3.2. Session 17.1. NKM-HRD Aspects of Newbuilds and Supporting Research

Session Chair: M. Skuce, Canada

IAEA Session Rapporteur: T. Reysset

The session explored critical themes in nuclear knowledge management (NKM) and human resources development (HRD) relevant to new nuclear power projects and research endeavours. Presentations covered strategies for preparing workforces, integrating management systems, preserving and transferring nuclear knowledge, and innovating in nuclear technology applications. These discussions emphasized enhancing competence, effectively managing resources, and fostering collaboration within the nuclear sector to support sustainable development in newbuild projects and research advancement.

Paper ID 107: Preparation of the HAEA for the licensing of new NPP units in Hungary

Speaker: G. Sárdy, Hungary

The presentation presents experience from the Hungarian Atomic Energy Authority (HAEA) in preparation for the licensing of the 2 new units in Paks NPP. The HAEA started preparing in 2009 by revising regulations and establishing a Project Department for overseeing the new NPPs. By 2014, HAEA specified required competencies and resources for this oversight. The presentation discusses HAEA's educational planning process, implementation, and key outcomes, including successes and challenges.

Paper ID 95: NKM and HR Challenges in Slovakia

Speaker: V. Szabó, Slovakia

The presentation provides an overview of Slovakia's current nuclear landscape, covering the completion of Units 3 and 4 at the Mochovce Nuclear Power Plant, ongoing operations at other nuclear facilities, nuclear waste management activities, and decommissioning processes at the Bohunice site. It will address challenges in human resource management within the nuclear sector abroad, as well as nuclear knowledge management both at nuclear power plants and within regulatory bodies.

Paper ID 115: Implementing Knowledge Management Programme in a Research and Development Organization – TENMAK

Speaker: A. Aylangan, Türkiye

The paper describes the Turkish Energy, Nuclear and Mineral Research Agency (TENMAK) approach to NKM. TENMAK implements robust management systems including information security, quality management, and innovation management. The NKM system is integrated with these systems for enhanced effectiveness and sustainability. Establishing an information management system is a key corporate strategy for the future.

Paper ID 264: Importance and Current State of Preservation Sharing and Management of Nuclear Knowledge at AERE of Bangladesh Atomic Energy Commission

Speaker: S. K. Chakraborty, Bangladesh

The paper discusses the NKM approach in the AERE of BAEC. Nuclear knowledge from R&D activities is preserved through annual publications of internal and technical reports. University students conduct research supervised by AERE scientists, contributing to thesis work and knowledge accumulation. Senior scientists impart their lifelong research experiences through discussion meetings and internal training programmes to transfer knowledge to younger generations. Collaborating with other institutions enhances opportunities for knowledge sharing and supports HRD initiatives.

Paper ID 79: Technology Roadmap for the Treatment and Reuse of Industrial Effluents: A Synergistic Approach for Advancing Research, Development and Innovation in Nuclear Science and Applications

Speaker: T. Campos, Brazil

The paper presents CNEN "Mobile Unit with Electron Beam Accelerator" project to enhance national capacity for treating industrial effluents using innovative electron beam technology. The project focuses on mapping essential competencies and infrastructure, forming balanced teams, and fostering collaborative relationships between Science and Technology Institutions and industry stakeholders. This initiative aims to advance nuclear technology R&D, promote personnel development, and align with broader sustainability and developmental goals in Science, Technology, and Innovation.

8.3.3. Session 17.2. Capacity Building for Safety and Security Including NKM-HRD Issues for Regulators and Their Licensees

Session Chair: L. Guo, IAEA

IAEA Session Rapporteur: M. Ovanes

The session focused on human resource development and training for regulators as well as nuclear safety knowledge management. The retention and motivation of practitioners for safety, in view of uncertainties and the rapidly changing political environment, was an overall identified challenge. National experiences highlighted the use the IAEA long-standing tool SARCoN (Systematic Assessment of Regulatory Competence Needs) as useful frameworks for establishing strategies to facilitate retention, staff jobs flexibility and promotion, also as a tool to structure the management of safety knowledge. Practical learning, on-the job training and mentoring as well as motivation of the staff were recognised as key elements for the development of human resources. International cooperation and the establishment of collaborative centres were found beneficial for capacity building at a national level. The use of e-learning as a knowledge base to support nuclear security programmes was highlighted.

Paper ID 130: Principles, Status and Plans for the Personnel Training of Nuclear and Radiation Safety Regulation in China

Speaker: W. Sheng, People's Republic of China

The Chinese National Nuclear Safety Administration (NNSA) oversees the workforce for nuclear safety. The presentation described the current training system of NNSA. Practical training, thorough qualification system and link to the training positions and potential promotions are the principles of the training. On-line training is available to all, complemented by external visits and forums. Mentoring, on the job daily training is conducted. Looking forward is focusses on motivation boost of new trainees and enhance of the overall system.

Paper ID 344: Competency Mapping, Career Path and SARCoN Assessment as Pivotal Tools for Staff Retention in Regulatory Bodies

Speaker: B. Bernard, Belgium

B. Bernard identified main challenges such as the retention and motivation of practitioners for safety, in view of uncertainties due to the rapidly changing political environment. He illustrated how SARCoN can be used to support some of these challenges. A strategic project is in place including three elements: 1. Competence mapping, 2. Identify positions that can be flexible or re-converted. 3. Career path that proposes flexibility, training to move horizontally and/or promotion.

Paper ID 199: Evaluation of National Approaches and Experience for Managing Nuclear Safety Knowledge in the United Republic of Tanzania

Speaker: D. Shao, Tanzania

D. Shao presented the process used to derive the specific competences from the regulatory functions. He underlined the benefit of having the role of a training coordinator and external cooperation. He presented the competence framework of TECDOC 1254, later superseded by safety report SRS 79 and SARCoN (TECDOC 1757). Using this 4 Quadrant model of competence framework, a knowledge base of hundreds of safety documents was produced. These reverted on more than 40 staff trained effectively and an evaluation programme.

Paper ID 253: Impact of Capacity Building on expertise of Pakistan Nuclear Regulatory Authority Workforce

Speaker: A. Murtaza, Pakistan

The presentation focussed on describing the PNRA's training system. It illustrated the recent establishment of a collaborative training centre with the IAEA. It highlighted the use of IAEA training materials such as the Basic Professional Training course, the SARCoN methodology and a new leadership programme, for which they also mentioned the IAEA School for Nuclear and Radiological Leadership for Safety. A. Murtaza underlined how the PNRA training system has proved effective, its continuous development and she underlined that PNRA has become an expert provider internationally.

Paper ID 435: Use of a Knowledge Base in IAEA Nuclear Security E-learning Programme: A Library of Resources for Programme Sustainability

Speaker: Y. Shin, IAEA

The presentation presented the e-learning resources, its development and multi linguicism approach. Several modules in the area of nuclear security were described as well as the process for using them. She underlined the role of e-learning for structuring knowledge and facilitating the efficiency of knowledge management and learning.

8.3.4. Session 17.3. Education & Training for Nuclear

Session Chair: Y. Wang, People's Republic of China

IAEA Session Rapporteur: H. Zhivitskaya

The session explored educational and capacity-building initiatives in nuclear technologies across various regions. Key topics included the Arab Atomic Energy Agency's (AAEA) efforts in advancing peaceful nuclear technologies and the educational impact of the Kindai University reactor in Japan. It highlighted the European Nuclear Experimental Educational Platform's (ENEEP) emphasis on hands-on experience, the IAEA's Nuclear Knowledge Management School at Texas A&M University, and global nuclear safeguards capacity building by the IAEA. Additionally, it showcased Rosatom's innovative training techniques, which personalize education and engage youth in nuclear projects. These initiatives collectively aim to enhance expertise, ensure safety, and foster the next generation of nuclear professionals.

Paper ID 267: Contribution of the Arab Atomic Energy Agency in HRD in Arab Countries: Three Decades of Experience

Speaker: K. Zahraman, Arab Atomic Energy Agency

The paper explores the Arab Atomic Energy Agency's (AAEA) mission to advance peaceful nuclear technologies in Arab countries through extensive capacity-building initiatives. Established in 1989, AAEA conducts over 30 training programmes annually, focusing on various sectors like health, environment, and nuclear power. These programmes aim to enhance expertise, transfer knowledge, and develop human resources in nuclear technologies. Through partnerships with international bodies like the IAEA, AAEA strengthens regional capabilities, particularly in radiation safety and nuclear security, ensuring sustained progress in nuclear science and applications across the Arab world.

Paper ID 108: The Expanding Role of the Kindai University Reactor as a Powerful Educational Tool for Nuclear Human Resource Development

Speaker: G. Wakabayashi, Japan

The paper discusses the educational impact of the Kindai University reactor, UTR-KINKI, a low-power research reactor crucial for nuclear human resource development in Japan. Since 1961, UTR-KINKI has served as a practical training tool for university nuclear engineering programmes and outreach to secondary education. Supported by a government consortium, it facilitates educational activities and international training workshops, fostering interest in nuclear science careers. This presentation highlights the reactor's role in shaping future nuclear professionals and emphasizes the effectiveness of low-power reactors in educational contexts.

Paper ID 402: ENEEP – Hands-on Education and Training for Nuclear Industry

Speaker: J. Lüley, Slovak Republic

The paper explores the pivotal role of experimental education at research nuclear reactors in enhancing nuclear science and engineering training. While computer modelling advances, practical hands-on experience remains crucial. The European Nuclear Experimental Educational Platform (ENEEP) exemplifies this approach, linking research reactors and specialized laboratories across Europe. Founding members, including institutions like Jožef Stefan Institute and Budapest University of Technology, prioritize experimental education to build competence. This paper emphasizes the irreplaceable value of in-person attendance for nuclear facility visits, hands-on experimentation, and direct interaction with instructors, highlighting the unique benefits not fully captured by remote education activities.

Paper ID 195: The IAEA Nuclear Knowledge Management School at Texas A&M University: Implementation Review and Lessons Learned

Speaker: K. Ragusa, United States of America

The paper explores the implementation of the IAEA Nuclear Knowledge Management (NKM) School hosted by the Centre for Nuclear Security Science and Policy Initiatives (NSSPI) at Texas A&M University in 2022 and 2023. This multinational event at the largest nuclear engineering education programme in the US featured active learning sessions developed in collaboration with IAEA experts and local industry leaders. The presentation details the course structure, participant demographics, feedback, and lessons learned, highlighting outcomes and adjustments made over successive years to enhance the NKM school's effectiveness.

Presentation: Nuclear Safeguards Capacity Building by the International Atomic Energy Agency

Speaker: J. Rahim, IAEA

The presentation discusses Nuclear Safeguards Capacity Building by the International Atomic Energy Agency (IAEA). It explores initiatives aimed at enhancing global nuclear security through training, technical assistance, and collaborative partnerships. Highlighting the importance of safeguarding nuclear materials and facilities, the presentation emphasizes the role of capacity-building programmes in promoting international cooperation and compliance with nuclear non-proliferation obligations. Case studies and outcomes illustrate the effectiveness of these efforts in strengthening safeguards implementation worldwide.

Paper ID 301: Dual Education for Engineers and Support Manpower for Nuclear Industry and Technology

Speaker: L. Lebedeva, Russian Federation

The presentation presents Rosatom's innovative training and education support for nuclear industrial and technological projects. Utilizing unique training techniques and a comprehensive testing system, Rosatom enhances the education process with detailed data analytics on students and trainers. This approach, surpassing the Vision Zero concept, personalizes educational pathways, engages youth in national nuclear projects through professional tests and vocational guidance, and emphasizes individual autonomy. It highlights Rosatom's successful practices in supporting initiatives from schoolchildren to young professionals, addressing competency gaps, and fostering the development of a new generation of engineers.

8.3.5. Session 18.1. Empowering Healthcare Through NKM and HRD

Session Chair: N. Bonilla, OECD-NEA

IAEA Session Rapporteur: J. Roberts

The session focused on the expansion of nuclear medicine, addressing the need for increased availability of facilities, workforce planning, and maintaining high standards of quality and safety. It highlighted challenges related to human resources, including the concentration of services in specific areas and the need for workforce renewal and adaptation to technological advancements. Emphasis was placed on the importance of extensive training and certification for medical physicists, as well as the role of staffing tools and quality management systems in ensuring effective and safe nuclear medicine practices.

Paper ID 211: Genesis of KM in a Network of Nuclear Medicine and Radiotherapy Centres in Developing Country

Speaker: V.J. Ugarte Ferrero, Argentina

The paper summarized the experience gained by the National Atomic Energy Commission of Argentina over the past years and how they plan to adopt a KM programme in a complex scenario of multiple networked organizations dedicated to nuclear medicine and radiation therapy, in the context of developing countries with ever-changing economic and political settings, scarce human resources and intensive use of technology.

Paper ID 84: The Process Approach in Human Resources Management in Nuclear Medicine

Speaker: F. Pérez González, Cuba

The work presents the experience of application for five years of an approach focus on the adaptation of recent graduate and of new workers to Nuclear Medicine work environment related to on-the-job training to obtain the corresponding Individual License issued by Regulatory Authority. Based on the evaluation of the performance of the personnel benefited by this approach, it can be concluded that although there remains a tendency towards instability of the workforce, the indicators of work quality and the flexibility to assimilate new technologies are higher than shown by technologists and specialists trained without this approach.

Paper ID 369: Medical Physics MSc Education in Hungary

Speaker: C. Pesznyak, Hungary

The paper presents medical physics MSc programme model in Hungary. The course curriculum comprises fundamental physical subjects (atomic and molecular physics, nuclear physics, particle physics) as well as fundamental medical knowledge (anatomy, physiology, radiobiology) required for subjects of diagnostics and therapy. The goal of the MSc education is for the students to be able to meet the special requirements of medical physics with the knowledge they have acquired, with professional awareness and compliance with ethical standards.

Presentation: A Model to Assess Staffing Needs in Nuclear Medicine: An IAEA Tool

Speaker: A. Brink, IAEA

A. Brink described the IAEA tool to assess staffing needs in nuclear medicine. Diagnosis has developed from visible screening to molecular screening. Theragnostic involves the combination of diagnosis and therapy. There is a paucity of qualified professionals in low- and middle-income countries but there is no point training significantly more if the availability of facilities is very low as is currently the case. A staffing calculator for nuclear medicine department staff has been developed by the IAEA. There is a separate medical physics staffing calculator. Quality management in nuclear medicine is maintained through QUANUM audits. There is also a similar system for radiology – QUAADRIL, Quality Improvement Quality Assurance Audit for Diagnostic Radiology Improvement and Learning.

8.3.6. Session 18.2. NKM and HRD Considerations in the Development of SMRs Technologies and their Applications

Session Chair: Y. Sun, People's Republic of China

IAEA Session Rapporteur: M. Ovanes

As more industry groups and governments are looking at nuclear to play an essential role in meeting climate and energy security goals, recent trends for capacity building in the nuclear industry have seen a significant increase in international cooperation and partnerships for the development of the nuclear supply chain and its workforce. This includes Member States exploring options to better and share knowledge, capabilities and resources in key strategic areas that will advance the small modular reactors and advanced reactors technologies and accelerate their deployment by 2030.

The China-Indonesia Joint Laboratory on High Temperature Gas-cooled Reactor exemplifies such partnership that focuses on joint research, training, technology transfer to enhance scientific and technological capabilities in both countries. The partnership is indicative of a resolute longer-term commitment to strengthen scientific and technological ties and promote sustainable nuclear power development through shared resources and knowledge transfers.

Rosatom's corporate strategy vision for 2023 contributes to building and sustaining capacity in nuclear organizations to support national nuclear programmes and ensure a skilled workforce for the future. The strategy includes a systematic approach to address challenges in human resource development, particularly for SMR NPP projects, amidst high market competition and understaffing. Initiatives such as collaboration with engineering schools, competency centres, and digital tools for education are highlighted as successful examples of human resource development, particularly when extended to external partners to support sustainable development of advanced nuclear technologies.

Paper ID 463: China-Indonesia Joint Laboratory on HTGR in the Context of NKM and HRD

Speaker: Y. Sun, People's Republic of China

The paper outlines the establishment of the China-Indonesia Joint Laboratory on High-Temperature Gas-Cooled Reactors (CI-JL-HTGR), a collaborative effort between Tsinghua University in China and Indonesia's National Atomic Energy Agency, aimed to accelerate research, foster talent and promote sustainable nuclear energy solutions by advancing the development of HTGR technology. The partnership contributes to strengthen the nuclear knowledge management and human resource development, pool the expertise, resources and the research capabilities to advance HTGR technology. Key activities include training workshops on reactor design and simulation tools, joint seminars for project development, and collaborative research on modular HTGR features, safety, and localization. The joint collaboration supports Indonesia's nuclear energy ambitions and fosters long-term scientific cooperation and capacity building in HTGR technologies between the two countries.

Paper ID 346: Personnel Training for New Business Solutions: The Case of SMR Nuclear Power Plants

Speaker: S. Pikh, Russian Federation

The presentation discusses Rosatom's Corporate Strategy Vision for 2030, focusing on enhancing knowledge and human resources' potential. It highlights a systematic approach to personnel training for new business solutions, particularly in Small Modular Reactor (SMR) nuclear power plants. The presentation showcases successful methods such as collaborations with engineering schools, internal competency centres, and digital tools for self-development. Additionally, it emphasizes the continuous improvement of both soft and hard skills and the exchange of best practices within Rosatom and with external partners.

Presentation: NKM-HRD Considerations in the Development of SMR in Indonesia: Case of PeLUIt-40

Speaker: N. Trianti, Indonesia

The presentation discusses Indonesia's efforts in capacity building as it pursues small modular reactor (SMR) technology to achieve net-zero emissions by 2060. The PeLUIt-40 reactor, a high-temperature gas-cooled SMR designed for electricity and industrial steam, is central to Indonesia's energy transition strategy. Indonesia's history with research reactors and its roadmap for emission reduction through technologies such as nuclear energy, hydrogen, and carbon capture are highlighted. Emphasis is placed on international collaboration and partnerships to build competencies across sectors, strengthen nuclear knowledge, and integrate SMRs into the national energy mix.

Presentation: Advances in Technology Developments of Small Modular Reactors and Microreactors

Speaker: A. Constantin, IAEA

The IAEA offers a multitude of resources and platforms to facilitate collaboration, knowledge capture and dissemination on the development and deployment of advanced reactor technologies, including Small Modular Reactors (SMRs).

IAEA's broad range of activities on SMRs span across technology development and deployment, economics, safeguards-by-design, commissioning and operation approaches, fuel cycle, waste management and decommissioning, infrastructure development, safety and security, industrial harmonization and standardization, and reactor technology assessment.

Comprehensive publications – such as the IAEA SMR ARIS Booklet offering descriptions of 83 different reactor designs – provide the current state-of-knowledge and highlight key SMR attributes such as economics, modularization, flexible application, smaller footprint, and potential integration with renewables. The IAEA's SMR Platform supports coordination of SMR activities and serves as a focal point for Member States for assistance on SMR-related issues. Additionally, a dedicated Technical Working Group on SMRs, composed of renowned international experts, advises on IAEA's programmatic activities on SMRs.

These initiatives highlight IAEA's commitment to fostering international collaboration and knowledge transfer contributing significantly to the advancement and deployment of SMR technologies and strengthening evidence-informed decision-making within the IAEA and Member States.

Presentation: Safety Demonstration for innovative reactors (SMRs, NWCRs)

Speaker: S. Poghosyan, IAEA

The presentation discusses the IAEA activities on safety demonstration for innovative reactors such as SMRs and non-water-cooled reactors (NWCR). It describes the recent and ongoing work on the publications in this area, in particular the review of applicability of IAEA Safety Standards to SMRs and NWCRs and relevant safety guide and safety reports being developed on safety assessment (e.g. DSA, PSA) for these reactors. It also describes the support offered by IAEA to the Member States planning to deploy SMRs or NWCRs such as the IAEA Technical Review Services and capacity building activities in this area.

8.3.7. Session 18.3. Modern Tools in NKM and HRD, Part 2

Session Chair: A. Di Trapani, Italy

IAEA Session Rapporteur: R. Kvetonova

The session continued the discussion from Session 10.1. and another new topic relating to the modern tools in HRD and NKM have been introduced. It has been highlighted the importance of early investment in nuclear leadership development, which can address human resource and knowledge management challenges. By stating that the long-term success of a nuclear power organization depends on encouraging leadership development. International Nuclear Information System (INIS) and its role in facilitating the exchange of information related to the peaceful use of nuclear energy has been introduced. The session presented an in-depth analysis of the data, showing an increasing interest in INIS. Topics focusing on improving cyber incident response in nuclear facilities by proactively detecting and preventing cyber threats have been also discussed. Use of full scope simulators, benefits of digital panels, including better Key Performance Indicators, application of knowledge management methodology to enhance nuclear science programmes, development and application of tools and modules with WOW Factor for School Education have been introduced and discussed too.

Paper ID 309: The Need for Nuclear Leadership Development Programme at the Early Stage of a Nuclear Power Programme: Ghana's Experience as a Newcomer Country

Speaker: C.K. Klutse, Ghana

The paper discusses Ghana's journey in developing a national nuclear leadership programme to support its nuclear infrastructure development. As a newcomer to the nuclear industry, Ghana faces the challenge of developing effective leadership for the safe and efficient implementation of its nuclear power programme. The paper details the steps followed in developing the nuclear leadership programme, identifies some of the challenges that were met, and suggests opportunities for improvement. It emphasizes the importance of early investment in nuclear leadership development, which can address human resource and knowledge management challenges. The paper concludes by stating that the long-term success of a nuclear power organization depends on encouraging leadership development from the outset.

Paper ID 138: The International Nuclear Information System (INIS): Trends and Analysis

Speaker: V. P. Nguyen, IAEA

The paper discusses the International Nuclear Information System (INIS) and its role in facilitating the exchange of information related to the peaceful use of nuclear energy. Established in 1969, INIS has collected over 4.4 million bibliographic records, with more than 2 million being full-text documents. The paper presents an in-depth analysis of the data, showing an increasing interest in INIS, especially from countries expanding their nuclear power programmes. The author also provides technical and policy recommendations to optimize the operation of the INIS Repository. The paper further explores the user profile of INIS, noting a marked increase in public interest since its database was made publicly accessible. Lastly, the paper highlights the need for INIS to broaden its scope to climate-related subjects, given the close linkage between nuclear energy and climate change. The author concludes by emphasizing the need for INIS to reach out to the public, especially in countries where English is not the native language, and to harvest more records in languages other than English from lesser-exposed institutions and sources.

Paper ID 64: Online Knowledge Opportunism to Nuclear Knowledge Management Strategies: Implications for Nuclear Cyber and Computer Incident Response Team

Speaker: D. Hossain, Bangladesh

The paper discusses the integration of online knowledge opportunism into nuclear knowledge management strategies. The focus is on improving cyber incident response in nuclear facilities by proactively detecting and preventing cyber threats. The paper introduces a framework called OKO2NKMS (Online Knowledge Opportunism to Nuclear Knowledge Management Strategies) which is designed to help nuclear cyber incident response teams update their Cyber Design Basis Threat (CDBT). The framework suggests extracting knowledge from unconventional online sources like hacker forums and Darknet, which often contain valuable information about emerging threats. The paper concludes by emphasizing the importance of this approach in building a sustainable cyber defence system for national nuclear programmes and other radiation facilities.

Paper ID 198: Digital Full Scope Simulators: Upskilling the Nuclear Workforce Prior to Reactors' Modifications and Upgrade

Speaker: S. Michelot, France

The paper discusses the use of Full-Scope Simulator (FSS) control rooms equipped with digital programmable panels in nuclear power plants. These panels can simulate different reactors by loading configuration files, allowing staff from different reactors to be trained at the same FSS. This technology addresses challenges in Nuclear Knowledge Management (NKM) and Human Resource Development (HRD), particularly for plant staff operating reactors of the same design but trained at a unique FSS. The digital panels make changes in the FSS control room easy, inexpensive, and quick, improving training conditions for plant staff and trainers. The paper also discusses the benefits of these digital panels, including better Key Performance Indicators (KPI), easy integration of plant design modifications, and flexibility. However, it also acknowledges the challenges, such as potential procurement issues, troubleshooting, user acceptability, and licensing of MCR staff. The paper concludes by discussing the development of next-gen compact simulators for Just-In-Time training sessions for MCR staff.

Paper ID 380: Applying Knowledge Management methodology to enhance Nuclear Science Programmes

Speaker: I. M. Paponetti, Italy

The paper discusses the application of knowledge management methodology to enhance nuclear science programmes. The authors propose a systematic approach based on Semantic Web technologies to create a Knowledge Base (KB) system for collecting and linking distributed information. This approach is extended to the nuclear domain through a foundational conceptual framework based on the Elementary Multiperspective Materials Ontology (EMMO). The paper also presents a practical application case developed within a university teaching environment, involving master students guided by field experts. The authors conclude by emphasizing the potential of this approach in environments where disparate information systems operate autonomously but necessitate integration for specific use cases, applications, or shared objectives, such as the nuclear energy field.

Paper ID 232: Development and Application of Tools and Modules with WOW Factor for NST School Education

Speaker: T. Iimoto, Japan

The paper discusses the development and application of educational tools and modules for nuclear science and technology (NST) education in schools. The authors have developed several tools such as a cloud chamber, a next-generation environmental radiation survey meter, an assembling kit for radiation counting, radiation sources made of natural substances, and NST educational movies. These tools aim to enhance the learning experience and sustain students' motivation. The authors have received positive feedback from pilot activities in various countries. They believe that the discussion on how to obtain and apply these educational tools effectively and safely in school classes should be international. They also suggest that the International Atomic Energy Agency (IAEA) should provide guidelines for manufacturing and using radiation sources for educational purposes. The authors conclude by emphasizing the need for skilled experts to support schools and teachers in NST education.

9. CLOSING SESSION

Closing statement and President's Report as provided, verbatim.

E. Pule
Conference President and HR Executive of Eskom Holdings SOC Ltd, South Africa

Excellencies, ladies and gentlemen,

After five days of intensive discussion, we are approaching the end of the International Conference on Nuclear Knowledge Management and Human Resources Development. I would like to express my sincere gratitude to all speakers, panellists, and participants who have joined us, here in Vienna and virtually, for your valuable contributions to this conference organized by the IAEA.

Nuclear knowledge management and human resources development have proven to be significant levers and opportunities for enhancing the effectiveness and sustainability of the global nuclear sector, due to their dynamic nature and the current evolving global landscape.

Over the past week, our conversations have been structured around key themes: people, technology, alliances, and sustainability. This approach has allowed us to share our expertise from diverse perspectives and think of innovation and growth through co-creation of collective knowledge.

By valuing diverse human viewpoints and leveraging cutting-edge technologies, such as AI, we can build strong collaborations, essential for a sustainable development and long-term success. The enthusiastic and active participation we have witnessed at this conference is a testament to this collaborative spirit.

At this point, let me remind you that we have had more than 760 participants from 107 Member States and 9 invited International Organizations in attendance this week. We received many paper submissions, which were presented orally and as posters or e-posters.

This week has been bustling with activity – 15 high-level plenaries and 24 parallel technical sessions were conducted, in addition to 8 side events featuring various topics organized by Member States, the nuclear industry, NGOs, and the IAEA Secretariat.

A great number of senior officials from IAEA Member States, nuclear industry leaders, international experts, and relevant stakeholders were in attendance. They delivered remarkable speeches, contributed to technical presentations, actively participated in discussions, and presented their scientific studies and research.

Thank you all for your dedication and collaboration in making this event a remarkable success.

Now, allow me to summarize the key findings from each day of our conference.

On Day One, Member States shared "Holistic Views on Nuclear Energy Policies and Their Impact on Nuclear Knowledge Management (NKM) and Human Resources Development (HRD); Global Challenges and National Alternatives." We heard national statements from Argentina, China, Egypt, France, Ghana, Japan, Morocco, the Russian Federation, Hungary, Saudi Arabia, South Africa, the United Arab Emirates, and the United States of America. These statements underscored a shared commitment to support national capacity building to create a competent and capable workforce at both national and international levels. This joint effort, tailored to each nation's sovereign approach, highlights the uniqueness of the nuclear sector and the commitment to further developing human capital and nuclear knowledge worldwide. It reaffirms the IAEA's role as the global centre for cooperation in the nuclear field, promoting the safe, secure, and peaceful use of nuclear technology.

In our first plenary panel, we discussed strategies for retaining talented women in our nuclear organizations. The focus was on identifying opportunities and best practices to foster their career development. Emphasis was placed on the importance of supporting the development of leadership skills and involving men as allies to support women in the nuclear field.

On Day Two, we focused on people and our opportunities to develop, empower, and lead. Presentations in the plenary and parallel sessions covered various aspects of innovative training programmes, national nuclear workforce development programmes, and global competency models. We also discussed nuclear safety and security culture, knowledge transfer, and defining critical knowledge. Additionally, the nuclear knowledge management and human resource development (NKM-HRD) aspects of fusion organizations were addressed in a session and a side event hosted by EUROfusion. The day concluded with an exciting and inspiring discussion on nuclear leadership, followed by a bold side event on international mobility hosted by France.

On Day Three, we explored technology as a key lever for enabling innovations and knowledge. The sessions and discussions focused on technical areas such as artificial intelligence, digital platforms, and other innovative technologies supporting Nuclear Knowledge Management (NKM) and Human Resources Development (HRD). Artificial intelligence was recognized as a current disruptor in science and technology development, with direct implications for knowledge management and learning processes. We also covered issues related to establishing new NKM programmes and lessons learned from existing ones. The day included a side event hosted by the NKM Section of the IAEA and a truly remarkable evening side event hosted by the Russian Federation.

On Day Four, we emphasized the power of alliances, aiming at engaging youth through global collaboration. The day began with keynote addresses on human resource strategies and attracting young people to the nuclear industry. Plenary sessions discussed the empowerment of the next generation and human-centric approaches in nuclear education. University education sessions explored various international nuclear engineering programmes, while discussions on education networks and attracting young talent highlighted initiatives in competency building and talent acquisition. The afternoon featured sessions on communities of practice, outreach education programmes, and a moderated panel discussion, concluding with a side event on enhancing nuclear competence through European collaboration hosted by ENEN.

Today, on Day Five, we looked towards the future, discussing how to build together resilience and progress for the sustainability of our human and knowledge ecosystems. The day commenced with keynotes addressing the challenges and strategies for human resource development in the nuclear sector, followed by a moderated panel on sustainable stakeholder engagement for capacity building development. Parallel technical sessions covered topics such as NKM and HRD for new nuclear builds, regulatory safety and security, and educational training. Afternoon sessions explored the role of NKM and HRD in healthcare, the development of small modular reactor (SMR) technologies, and modern tools in NKM. The day also included an IAEA side event and a technical tour of the Nuclear Security Training and Demonstration Centre in Seibersdorf.

Ladies and gentlemen,

Let us reflect on the key takeaways from this conference and consider how we can implement them in the coming years. Leading, engaging, and developing human capital and nuclear knowledge will be the spearhead of our efforts moving forward.

As we strive to attract and retain top talent, the nuclear sector must position itself as an exciting and progressive scientific and industrial field. This involves effectively drawing in the brightest minds by showcasing our innovations and advancements.

Moreover, it is essential to acknowledge the values and motivations of young professionals today. Many are driven by a desire to contribute to long-term social and global goals such as sustainability and climate change mitigation. By emphasizing nuclear's contributions to these global objectives, we can attract young people who are passionate about making a positive impact on the world.

Nuclear must be seen as a modern, high-tech industry capable of utilising the latest technologies, such as AI, and captivating the labour market's attention, with a particular focus on youth. Building a culture that embraces diversity in terms of gender, age, and background will ensure equality and inclusion, enabling effective engagement of both national and international workforces.

Strengthening international cooperation and building strong partnerships among global organizations will facilitate the development and deployment of international workforces for international projects, reflecting the transnational nature of nuclear activities.

To remain competitive, the nuclear industry must proactively offer attractive student projects and open positions to students early in their education. Engaging students at this formative stage will help build a strong pipeline of future nuclear professionals.

In conclusion, by addressing these key areas, we can position the nuclear industry as a competitive and attractive career choice for young professionals. We must leverage our technological advancements, align with the values of the younger generation, engage students early, offer competitive compensation, and provide dynamic career opportunities. By doing so, we will secure the talent necessary to drive our industry forward and contribute to a sustainable future.

Ladies and gentlemen,

The past week has been intense and productive. Stakeholders from various sectors—government, industry, international organizations, non-governmental organizations, women's groups, and the young generation—came together to discuss the opportunities and challenges facing the nuclear sector. We proposed constructive solutions to further enhance the knowledge and capabilities of our global nuclear workforce. I would like to take this opportunity to express my heartfelt thanks to each and every one of you for your professional and insightful contributions, which have led to very fruitful outcomes for the conference.

I would also like to extend my gratitude to the Scientific Programme Committee for organizing the conference and evaluating and reviewing the scientific contributions. My thanks go to all chairs and co-chairs, moderators, and rapporteurs of plenary and parallel sessions for guiding and supporting the interactive presentations and discussions. Additionally, I appreciate everyone who contributed with papers, presentations, and posters.

Special thanks go to our scientific secretaries, Pedro Diéguez Porras and Alesia Iunikova, the Nuclear Knowledge Management Section team, as well as the IAEA colleagues from Conference Services, and all those in the Agency who have worked hard to make this conference a success.

Ladies and gentlemen, thank you all for your dedication and collaboration in making this event a remarkable joint achievement. Please also have a safe trip home.

Closing remarks as provided, verbatim.

M. Chudakov
Deputy Director General and Head of the Department of Nuclear Energy, IAEA

Ladies and gentlemen, dear colleagues,

Good afternoon! Let me begin by expressing my sincere gratitude for everyone who played a role in making this conference an informative and stimulating event.

To our Conference President Ms Elsie Pule, our Conference Vice Presidents Mr Mitsuru Uesaka and Ms Kim Pringle, all of our keynote speakers, chairs, and moderators, the proSTEM Challenge winners – thank you. And thank you to Mr Vitali Shumski, Mr Oszvald Glöckler, Ms Ruvi Perera, Mr Gonglin Li, Mr Tom Danaher, Ms Andrea Kamara, and all of our other colleagues from Conference Services and the Nuclear Knowledge Management team who worked diligently to make it a success. And of course, I must thank our Conference clerks and technicians who have helped keep everything on track and running smoothly.

Finally, let me offer my thanks to the programme committee as well as our scientific secretaries Mr Pedro Diéguez Porras and Ms Alesia Iunikova for organizing such an excellent – and timely – programme of events.

Ladies and gentlemen,

More than 760 participants from 108 Member States and 9 invited organizations registered for our conference. This is an impressive number that underscores the high level of interest in the topics presented here. Throughout the week, we observed a high level of engagement, with many compelling questions and in-depth discussions.

This conference comes at a pivotal moment for the nuclear power industry. In order for nuclear power to maximize its potential, there is much to be done. The IAEA foresees nuclear capacity more than doubling by 2050 in the high case scenario, and we estimate there will be more than 4 million professionals supporting the nuclear power industry by 2050. We also project that about one-third of the existing workforce will retire by 2033, and many people, perhaps more than one million, will need to enter the workforce by then to replace those retiring and help support nuclear power expansion. We must be proactive in meeting these growing needs.

Overcoming these hurdles will not be easy. But we have the tools, and we have the motivation to achieve great things, and I am confident we will succeed.

It is critical that we identify and develop talented individuals so that they may become valued assets to their teams for years to come. And we must pursue development approaches that centre diversity and inclusion, and not only because this is the equitable thing to do, but also because innovation thrives when more people are given opportunities.

We must also be flexible and adaptive in our approach to skills maintenance and development and make use of new and emerging technologies such as AI. AI has the potential to optimize many processes in our industry, from training to operations, and it is important that we effectively leverage these technological advances.

Ladies and gentlemen,

It has also been discussed this week that the nuclear industry faces competition for talent from other industries, particularly those focused on advanced technologies. Addressing this issue is a multifaceted endeavour, and one of the ways to attract capable young people is through engaging student challenges, like our proSTEM challenge. Another is to recruit students for open positions in the early stages of their higher education. It is also vital that the nuclear industry offers good compensation packages and develops long-term employee retention strategies. And always, knowledge management, transfer and retention practices must continue to be developed and implemented in order to maximize the sustainability of our industry.

While there is much work to be done, there are also many reasons to be optimistic. We have heard this week about the highly robust knowledge management and human resource development efforts ongoing throughout the world. Sharing these experiences, these lessons learned, here at this conference will be a boon for the industry, I am sure. When we collaborate, when we have a plan – there is nothing we cannot accomplish.

And for the IAEA's part, our staff will continue to provide the world-class assistance that has been a hallmark of our organization for decades. Our capacity building programmes have supported the development of thousands of professionals around the world. We have helped universities develop curricula, we have assisted Member States in improving their knowledge management practices, and, through our Marie Sklodowska-Curie Fellowship Programme and Lise Meitner Programme, we have made real progress in addressing the need for more women in the nuclear workforce. I am confident that our team will continue to deliver and adapt to this dynamic industry as we strive together towards our common goals.

Dear colleagues,

This has been an engaging, highly productive conference, I think you will agree. Once again, I thank you all for your enthusiastic participation and look forward to our next opportunity for engagement.

I wish you all a safe journey home and declare the Conference adjourned.

Thank you.

ANNEX I.

CONFERENCE STATISTICAL DATA

International Conference on Nuclear Knowledge Management and Human Resources
Development: Challenges and Opportunities

STATISTICAL REPORT

Location:	IAEA Headquarters, Vienna, Austria
Scientific Secretaries:	Pedro Diéguez Porras (NEPK) Alesia Iunikova (NEPK)
Scientific Support:	Oszvald Glöckler (NEPK) Vitali Shumski (NEPK) Ruvi Perera (NEPK)
Admin. Support:	Andrea Kamara (NEPK) Yuliya Tulubtsova (NEPK)
Conference Coordinator:	Tom Danaher (MTCD)

REGISTRATION/ATTENDANCE:

Total no. of official participants:	763 including invited persons
From Member States:	741 from 107 Member States
From Permanent Observers	2 from 1 Permanent Observer
From Organizations	20 from 9 Organizations
Thereof Female participants	317 (43%)
Thereof Male participants	446 (57%)
Total no. of invited persons:	70

PROGRAMMATIC INFORMATION

Structure	1 x Opening Session with 2 Opening Statements; 2 x Sessions with High-Level Statements; 4 x Keynote Sessions; 3 x Plenary Sessions; 4 x Moderated Panel Discussions; 24 x Parallel/Technical Track Sessions; 1 x Special Session on Women in Nuclear; 8 x Side Events; 7 x Poster Sessions and 1 x Closing Session;

VIRTUAL ATTENDANCE

App Users:	840

1 July 2024	
Live Stage 1	570 views

2 July 2024	
Live Stage 1	741 views
Live Stage 2	250 views
Live Stage 3	600 views

3 July 2024	
Live Stage 1	519 views
Live Stage 2	301 views
Live Stage 3	284 views

4 July 2024	
Live Stage 1	777 views
Live Stage 2	424 views
Live Stage 3	488 views

5 July 2024	
Live Stage 1	875 views
Live Stage 2	467 views
Live Stage 3	764 views

1) Total number of official participants: 763

➢ 741 from 107 Member States:

Albania	1	Libya	2
Algeria	2	Malaysia	25
Argentina	17	Mali	1
Armenia	3	Mauritania	2
Australia	3	Mongolia	1
Austria	1	Morocco	6
Azerbaijan	3	Mozambique	1
Bangladesh	6	Myanmar	2
Belarus	1	Namibia	1
Belgium	4	Nepal	2
Benin	2	Netherlands, Kingdom of the	5
Bosnia and Herzegovina	1	Nicaragua	1
Brazil	13	Niger	2
Brunei Darussalam	2	Nigeria	20
Bulgaria	6	North Macedonia	1
Burundi	1	Norway	3
Cambodia	6	Oman	1
Cameroon	1	Pakistan	12
Canada	9	Paraguay	3
Chad	1	Peru	1
China	83	Philippines	7
Comoros	2	Poland	8
Congo	3	Portugal	1
Croatia	4	Qatar	4
Cuba	2	Romania	4
Czech Republic	5	Russian Federation	72
Democratic Republic of the Congo	5	Rwanda	3
Ecuador	1	Saint Kitts and Nevis	2
Egypt	18	Saint Lucia	1
Estonia	1	Saudi Arabia	11
Ethiopia	4	Serbia	3
Finland	3	Slovakia	7
France	27	Slovenia	1
Georgia	2	South Africa	8
Germany	11	Spain	8
Ghana	33	Sri Lanka	2
Greece	1	Sudan	2
Hungary	11	Sweden	7
India	4	Switzerland	3
Indonesia	28	Syrian Arab Republic	2
Iran, Islamic Republic of	3	Tajikistan	3
Iraq	4	Thailand	5
Israel	1	Togo	1
Italy	5	Tunisia	5
Jamaica	4	Türkiye	9
Japan	11	Uganda	14
Jordan	5	United Arab Emirates	8

Kazakhstan	2	United Kingdom	14
Kenya	19	United Republic of Tanzania	6
Korea, Republic of	7	United States of America	16
Kuwait	1	Uzbekistan	2
Latvia	3	Zambia	6
Lesotho	1	Zimbabwe	6
Liberia	1		

> ➢ 2 from 1 Permanent Observer:

State of Palestine	2

> ➢ 20 from 9 Organizations:

Arab Atomic Energy Agency (AAEA)	2
Electric Power Research Institute (EPRI)	1
European Commission, Joint Research Centre (EC/JRC)	1
European Nuclear Education Network (ENEN)	3
European Nuclear Society (ENS)	4
European Technical Safety Organisations Network (ETSON)	1
European Commission (EC)	5
International Energy Agency (OECD/IEA)	1
OECD Nuclear Energy Agency (NEA)	2

ANNEX II.

LIST OF PAPERS

Indico ID No.	Author(s) or presenter	Member State/ Organization	Paper or Abstract Title	Session No.
9	M.P. Lopez	Argentina	Knowledge Loss Risk Analysis: Recognizing the Criticality of People's Abilities and Skills	6.1
15	M.V. Simonyan	Armenia	Knowledge Management and Human Resource Development at Nuclear and Radiation Safety Centre	10.2
25	U.S. Adam M. Ibrahum	Nigeria	An Assessment of the Risk of Nuclear Knowledge Loss at the Nigerian Research Reactor Facility	10.3
29	A. Kosilov	STAR-NET	STAR-NET Continues to Build Competencies on Peaceful Application of Nuclear Technology	13.2
40	A.N. Kosilov N.I. Geraskin V.V. Kharitonov Yu. N. Volkov E.G. Kulikov	Russian Federation	Experience in Implementation of INMA Nuclear Technology Management Master's Programme at National Research Nuclear University MEPhI +Course for University Master's Level Programmes	14.1
45	D.L.C. Archila	Brazil	Capabilities in Nuclear Sciences and Business Management: Combining Performance, Practices and Organizational Resources at the National Nuclear Energy Commission (CNEN) of Brazil for Promoting Innovation	10.2
46	Y. Nara T. Ohkura A. Harada F. Sakamoto Y. Ikuta Y. Nakano T. Sakaguchi	Japan	Fostering Nuclear Professionals and Inspiring the Younger Generation by JAEA: Good Practices in Japan and Asian Countries	13.1
52	D. H. Nugroho	Indonesia	Strengthening the Knowledge Management Enabler in Nuclear Computer Security Planning Response Actions of Nearly Autonomous NPP through Hidden Markov Model	6.2
55	Ž. Tomšić K. Trontl	Croatia	Critical Mass of Experts Required for a National Sustainable Nuclear Buildup – Croatian Case Study	9.2
58	S. El-Morshedy	Egypt	Nuclear Knowledge Management System for Egypt Research Reactors	10.3
62	G.F. Puglia	Argentina	NKM implementation as a foundation stone in a Decommissioning Plan	9.3
64	D. Hossain	Bangladesh	Online Knowledge Opportunism to Nuclear Knowledge Management Strategies: Implications for Nuclear Cyber and Computer Incident Response Team	18.3

Indico ID No.	Author(s) or presenter	Member State/ Organization	Paper or Abstract Title	Session No.
78	L. Uyigue A. Kuye	Nigeria	Nuclear Security Education and Training Capacity Development at the University of Port Harcourt: Outcomes and Prospects	6.2
79	T. Campos D. Archila	Brazil	Technology Roadmap for the Treatment and Reuse of Industrial Effluents: A Synergistic Approach for Advancing Research, Development and Innovation in Nuclear Science and Applications	17.1
84	F. Pérez González	Cuba	The Process Approach in Human Resources Management in Nuclear Medicine	18.1
92	P.V. Petkov A. Georgiev G. Rainovski	Bulgaria	Challenges in Development of Nuclear Technology Management Education Programme in Sofia University "St. Kliment Ohridski"	14.1
95	V. Szabó	Slovakia	NKM and HR Challenges in Slovakia	17.1
102	N. Jahan M. Malek Soner	Bangladesh	National Status of Nuclear Knowledge Management and Human Recourse Development of BAEC TRIGA Research Reactor (BTRR)	10.3
107	G. Sárdy	Hungary	Preparation of the HAEA for the licensing of new NPP units in Hungary	17.1
108	G. Wakabayashi	Japan	The Expanding Role of the Kindai University Reactor as a Powerful Educational Tool for Nuclear Human Resource Development	17.3
112	E. Belonohy	EUROfusion	The EUROfusion Knowledge Management and Human Resource Development Strategy	5.2
113	S. Pleslic E. Jovicic	Croatia	Opportunities and Challenges in the Development of National Nuclear Knowledge Management Programme	5.1
115	A Aylangan S. Ünlü	Türkiye	Implementing Knowledge Management Programme in a Research and Development Organization – TENMAK	17.1
129	J. Repussard C. Thomas R. Kaminska Y. Guntzburger D. Pierre	France	Leadership for Safety Education: A Key to Successful Integrated Management and Safety Culture Enhancement	6.2
130	W. Sheng	People's Republic of China	Principles, Status and Plans for the Personnel Training of Nuclear and Radiation Safety Regulation in China	17.2
134	N. Velichkova	Bulgaria	Growing the Next Generation Nuclear Talents in Bulgaria	13.3
138	V.P. Nguyen	IAEA	The International Nuclear Information System (INIS): Trends and Analysis	18.3
146	C. Schönfelder	ENEN	NUCLEATION – a Community of Practice on Vocational Training in Nuclear Domains	14.2

Indico ID No.	Author(s) or presenter	Member State/ Organization	Paper or Abstract Title	Session No.
153	H. Elsayed	Egypt	Implementation of Nuclear Knowledge Management Programme on a New Nuclear Facility	5.3
167	C. Chevereau	France	NKM: Fostering a sustainable culture of nuclear knowledge sharing through standardised, efficient, active, and impactful Nuclear Communities of Practice	14.2
171	L. O'Brien	United Kingdom	The Application of AI to Support Sustainable Nuclear Technology Development	8
172	M. Gant L. Begley	United States of America	Nuclear Materials Integration Lessons Learned in Nuclear Knowledge Management	6.2
173	H. Adnan N. Ngadiron	Malaysia	Building a Sustainable Nuclear Workforce: Strengthening the Foundation through National Curricula	14.3
174	Zh. Li L. Niu J. Qui J. Zeng	People's Republic of China	From Documents to Knowledge: Research on the Application of AI Technology in Documents Knowledge Mining of Nuclear Power Plant: A case study of Fuqing NPP in China	9.1
177	C. Kwisanga B. Safari	Rwanda	Establishing Nuclear and Radiological Science Educational Programmes in Rwanda: Challenges and Opportunities	13.1
178	O. Balthazard	France	Know-How Transfer Method	6.3
180	L.A. Piciaccia R. Basili D. Croce	Norway / Italy	The Impact of Generative AI in NKM and HRD. Operational Nuclear Knowledge Enhancement and Management for New and Expert Users	9.1
195	K.H. Ragusa C.M. Marianno J.W. Crenshaw	United States of America	The IAEA Nuclear Knowledge Management School at Texas A&M University: Implementation Review and Lessons Learned	17.3
198	S. Michelot	France	Digital Full Scope Simulators: Upskilling the Nuclear Workforce Prior to Reactors' Modifications and Upgrade	18.3
199	D.W. Shao P.W. Shirima	Tanzania	Evaluation of National Approaches and Experience for Managing Nuclear Safety Knowledge in the United Republic of Tanzania	17.2
210	F. Puyade	France	Andra's innovative Forward-Looking HR Approach for Cigéo Construction Phase	9.3
211	V.J. Ugarte Ferrero R. Barrachina I. E. Roccaro	Argentina	Genesis of KM in a Network of Nuclear Medicine and Radiotherapy Centres in Developing Country	18.1
214	M. Gladyshev C. Scherer A. Kosilov	IAEA	Fostering the Next Generation of Nuclear Energy Sustainability Champions	14.1

Indico ID No.	Author(s) or presenter	Member State/ Organization	Paper or Abstract Title	Session No.
217	C. K. Bright D. Smeatham	United Kingdom	The Role of Knowledge Management in the Development and Deployment of Innovation in the Nuclear Sector	6.3
224	M.A. Akpanowo Y.U. Idris L. Abdulrauf M.A. Dandawaki	Nigeria	The Challenges and Prospects of Human Resource Development and Nuclear Knowledge Management in the Nigerian Nuclear Regulatory Authority	9.3
225	S. Mouline T. Marfak S. Oukemeni	Morocco	Leadership and Culture for Safety and Security Basis of Regulatory Capacity Building: AMSSNuR Experience and Lessons Learned	6.2
227	M. Ono N. Ogiwara	Japan	Hitachi-GE Nuclear Energy, Ltd.'s Nuclear Knowledge Management for Transferring Technical Knowledge to Next Generation	6.1
229	Y. Wang A. Yu S. Yang H. Zheng	People's Republic of China	Talent Cultivation for Global Nuclear Energy Development: Contributions and Practices from China	5.3
232	T. Iimoto R. Takaki T. Toda T. Kato T. Kakefu	Japan	Development and Application of Tools and Modules with WOW Factor for NST School Education	18.3
230	C.Yang	People's Republic of China	China National Nuclear Corporation's Innovative Practices in Human Resource Management: Human Resource Management System Standards and Quantitative Appraisal Method for Management Departments	3
234	S.J. Naumer P.T. Kekäläinen M. Dubovik E. Holt	Finland	Transferring knowledge of new LILW pre-disposal practices via case studies from the Euratom PREDIS project	9.3
242	T. Kita H. Morino T. Fujiwara	Japan	Empowering Japan's Nuclear Future – Strategic Initiatives in Talent Acquisition for the Nuclear Industry	13.3
249	A. Sa'id S.A. Jonah N. N. Garba	Nigeria	Nuclear Knowledge Management: The role of Centre for Energy Research and Training Zaria, in Education and Training of a Nuclear Workforce	5.3
253	A. Murtaza A. Danyal	Pakistan	Impact of Capacity Building on Expertise of PNRA Workforce	17.2
254	F. Lemont V. Morar V. Testard A. Magouh	France	INVICTUS: A shared multimodal project serving human resources development	10.1

Indico ID No.	Author(s) or presenter	Member State/ Organization	Paper or Abstract Title	Session No.
	D. Hertel F. Cames A. Vuignier			
255	A. Piagentini	European Commission	EU Nuclear Decommissioning Knowledge Management Initiative	9.3
257	V. Mitinskaya	Russian Federation	Intercultural Communications as an Element of NKM: Language Barriers and Ways to Overcome Them	6.1
258	V. Mazepov M. Solovieva	Russian Federation	Rosatom Technical Academy Experience of Organizing Training Events under the IAEA's Technical Cooperation Programme	5.3
260	Z. Bacs	Hungary	Multi-step NKM and HRD Process in the Paks 2 Project	6.1
261	P. Terry	United States of America	Improving Nuclear Industry Efficiency Through Neuroeducation and Neuroleadership	10.1
263	D. Mashkov V. Palamarchuk D. Podoliakin	Russian Federation	Personnel Training for First NPP in Country	5.1
264	S.K.Chakraborty	Bangladesh	Importance and Current State of Preservation Sharing and Management of Nuclear Knowledge at AERE of Bangladesh Atomic Energy Commission	17.1
267	K. Zahraman D. Mosbah	Arab Atomic Energy Agency	Contribution of the Arab Atomic Energy Agency in HRD in Arab Countries: Three Decades of Experience	17.3
269	G.L. Pavel K. Piliuhina	ENEN	ENEN's Contribution to the Creation of a Competent Nuclear Workforce	13.2
272	M. Al'Azri	Oman	Inaugural International Nuclear Science Olympiad: Fostering Excellence in Youth Education	14.3
277	K. S. Kang Ch. H. Kim	Republic of Korea	Implementation of Mentoring Programme to Facilitate Knowledge Management and Technical Transfer	6.3
280	E. Bologov N. Zhdanova L. Nikolaev	Russian Federation	Practice Oriented training of Foreign Specialists in Nuclear Security at the Rosatom Technical Academy	5.3
281	M. Coeck	Belgium	Nuclear Competence Building by the Belgian SCK CEN Academy	14.3
298	I. Cvetkov	Bulgaria	Approach to identify and manage critical knowledge at Kozloduy NPP	6.1
301	A.D.Ponomarenko	Russian Federation	Dual Education for Engineers and Support Manpower for Nuclear Industry and Technology	17.3
304	S. Gezer	Türkiye	The Nuclear Knowledge Management Necessity for Sustainable Energy Development	9.2

Indico ID No.	Author(s) or presenter	Member State/ Organization	Paper or Abstract Title	Session No.
309	C. K. Klutse C. Z. Dompreh Y. G.K. Mawugbe R. D. K. Ameevor A. Nyamful	Ghana	The Need for Nuclear Leadership Development Programme at the Early Stage of a Nuclear Power Programme: Ghana's Experience as a Newcomer Country	18.3
310	Z. J. Belmonte T. Iimoto A.L. Nery P. E. V. Lira C. Yu Md. M. Hasan	Philippines	Examining Key Factors Shaping Youth's Intentions in Pursuing Nuclear-Related Courses: Attitude Analysis for Workforce Sustainability	14.3
315	C. Pesznyak	Hungary	ENEN2plus – building European Nuclear Competence through Continuous Advanced and Structured E&T Actions	12
316	T. Gelo	Croatia	The Importance of Education and Training in Operation of a Nuclear Power Plant Krško	5.3
319	E.E. Rakhmankina	Russian Federation	System of Attracting and Development Scientific Personnel for Advanced Technology	14.3
320	G. Hoefer P. L. Wellmann	Germany	KM Concepts and Approaches in the German Waste Management Organisation (BGE)	10.2
322	A. C. Lopez	Spain	NEXA: Digitalization. Blended training approach. Standardization of curriculum	10.1
324	H. Malik	United Kingdom	Enabling Tacit Knowledge Transfer through Communities of Practices and Retention of Critical Knowledge in NWS	14.2
325	D. Nyarko S. Yamoah	Ghana	Knowledge Management in a Developing Country: The Case Study of Ghana	9.2
333	L. Cizelj A. Cagnac M. Coeck C. Creze F.J.Elorza Tenreiro G.L. Pavel C. Pesznyak T. R. Vollmer J. Starflinger F. Tuomisto D. Visvikis	Slovenia	Mobility in the Development of Nuclear Talents: Why and How	13.1
338	F. Ruiz Martinez R. Null	United States of America	Knowledge Management and the Global Competency Model at Westinghouse	5.1
344	B. Bernard M. Roobaert	Belgium	Competency Mapping, Career Path and SARCoN Assessment as Pivotal Tools for Staff Retention in Regulatory Bodies	17.2

Indico ID No.	Author(s) or presenter	Member State/ Organization	Paper or Abstract Title	Session No.
346	S. Pikh	Russian Federation	Personnel Training for New Business Solutions: The Case of SMR Nuclear Power Plants	18.2
347	A. Delmotte	Spain	AP1000/AP300 Design Training Model and Staffing Optimization	9.1
357	V. Karezin E. Nagibina	Russian Federation	ROSATOM International Education Network	13.1
360	O.N. Karmishina	Russian Federation	Digital HR-services for Nuclear Industry	10.1
364	V. Tsygoda N. Lobanova	Russian Federation	The synergy of data-centric and human-centric approaches within application of AI technologies for NPP digital modelling	9.1
369	C. Pesznyak A. Aszód Sz. Czifrus D. Legrády T. Major G. Stelczer M. Szieberth	Hungary	Medical Physics MSc Education in Hungary	18.1
370	M. Szieberth A. Aszódi A. Nemeslaki Sz. Czifrus	Hungary	Lessons Learned from Nuclear Technology Management Education at BME and Future Development Plans	14.1
373	A.H. Tkaczyk	Estonia	Development of a nuclear education strategy to meet human resource needs in Estonia	13.1
376	E. T. Zege	Ethiopia	The Challenge of Knowledge Management for Preserving Radiation Protection Capabilities in Ethiopia Technology Authority	6.3
380	I.M. Paponetti E. Ghedini F. Zaccarini A. Anchini S. Baldini S. Quaranta D. Rimondini	Italy	Applying Knowledge Management methodology to enhance Nuclear Science Programmes	18.3
382	M. Coeck S.G. Eeckhout A-L. Kochuyt E. De Roeck	Belgium	Tabloo: Informing Pupils, Students and the Public about Nuclear Science and Technology	13.3
384	A. Siebert L. Aldave De Las Heras K. Boboridis E. Colineau R. Eloirdi E. Michailidou	European Commission	A European perspective – The Nuclear Research Infrastructures Open Access Scheme of the Joint Research Centre (JRC) at the European Commission – Contributions to Education, Training, Mobility and Scientific excellence	14.2

Indico ID No.	Author(s) or presenter	Member State/ Organization	Paper or Abstract Title	Session No.
	R. Raison C. Fontana M. Hult C. Paradela P. Schillebeeckx K-L. Nilsson L. Iglesias Perez A. Jenet F. Taucer			
386	B. Eriksen M. Goulart M. Martin Ramos A. Eriksson	European Commission	European Human Resources Observatory in the Nuclear Field – A Status Update	10.2
387	C. Regnaud	France	How EDF Nuclear Generation Division Governance Creates Favourable Conditions to Develop and Promote Knowledge and Skills Sharing Amongst the Nuclear Fleet	14.2
388	J. Sowagi	Canada	Developing Nuclear Talent: A Strategic Approach to Tackling Workforce Challenges in the Nuclear Power Industry through Training	7
390	L. Iglesias Pérez	European Commission	The Foresight Strategy Applied to the Nuclear Knowledge Management	10.2
393	I. Abdul Rahman T. Jiejuan M.B. Mishar	Malaysia	The International Nuclear Science and Technology Academy (INSTA): A New Initiative in Supporting Regional Nuclear Knowledge Management and Human Resource Development	13.2
394	J. Porsmyr R. Szőke I. Szőke M. Jarallah	Norway	Innovative Digital Approach to Nuclear Fostering Education, Knowledge Management, and Workforce Development	9.1
395	P. Carbol N. Belmans M. Coeck A. Valls E. Holt A. Tatomir J. Faltejsek T. Beattie	European Commission	Lessons Learned of Knowledge Management Activities in EURAD and PREDIS	9.2
396	A. Zúñiga	Peru	Knowledge Management at Nuclear Reactor División in RP10	10.3
399	J. Najder C. A. Vega	European Nuclear Society	Impact of the Role of Nuclear Energy in Climate Topics on Attraction and Retention of Talents in Nuclear Sector	13.3

Indico ID No.	Author(s) or presenter	Member State/ Organization	Paper or Abstract Title	Session No.
401	F. Cantargi C. Gho G. Quintana G. Ferenaz M. Margutti M. Chocron	Argentina	CEATEN: the Postgraduate Course that Brings Nuclear Knowledge to Professionals from all Branches of Science and Engineering	14.3
402	J. Lüley V. Radulović	Slovak Republic	ENEEP – Hands-on Education and Training for Nuclear Industry	17.3
404	S. Ariyanto A.E.L. Conjares Y. Nam M. Mishar H. Zhivitskaya P. Pengvanich X. Guo Y. Kasesaz R. Dowler	Indonesia	ANENT's 20 Years of Learning and Sharing Nuclear Knowledge: A Historic Milestone	13.2
407	C. Hartnack A. Hartnack A. Abdelouas	France	The SARENA Program: Education in Nuclear Engineering within an European Mobility Scheme – Feedback from Experience and Future Perspectives	13.1
409	D.C. Invernizzi	United Kingdom	How Can a Connectivism-Based Approach Enhance Adult Learning and KM in Nuclear Decommissioning?	9.3
411	V. Barrins	Australia	General Manager People	13.3
412	S. Johnstone D. Rourke	United Kingdom	Revolutionising Nuclear Decommissioning with AI-enhanced KM: A Real-World Case Study	9.1
415	V. Richet	France	CurieLM: A Generative AI for Smart Knowledge Management in the Nuclear Industry	8
425	G.O. Balhamar	United Arab Emirates	E-learning Onboarding programme: "Iqrab W Ta'alam"	10.1
427	G. O. Balhamar	United Arab Emirates	Irshaad Youth Program	6.3
429	J. Gustafsson	Sweden	From strategic to operational competence planning and processes	5.1
435	Y. Shin M. Labyntseva	IAEA	Use of a Knowledge Base in IAEA Nuclear Security E-learning Programme: A Library of Resources for Programme Sustainability	17.2
438	S. Monti	European Nuclear Society	Ensuring Sustainable Nuclear Workforce Development in Europe and Beyond	9.2
443	I. Charit	United States of America	The INMA-Endorsed Nuclear Technology Management Program at the University of Idaho–Idaho Falls	14.1

Indico ID No.	Author(s) or presenter	Member State/ Organization	Paper or Abstract Title	Session No.
444	N. Hashim F. Omonya Wanjala	Kenya	The African Network on Education in Nuclear Science and Technology – AFRA-NEST	13.2
450	D.M. Tucker J.M. Hopwood	Canada	Developing a National Nuclear Workforce Development Framework in Canada	5.1
455	G. Chen X. Liu	People's Republic of China	Promoting High Speed Development of Nuclear Power through Standardized Human Resources Management System	6.1
452	A.M. Hooker J. McEvoy-May N.A. Spooner	Australia	Nuclear Education and Knowledge Management Activities at the University of Adelaide, Australia	14.1
463	Y. Sun	People's Republic of China	China-Indonesia Joint Laboratory on HTGR in the Context of NKM and HRD	18.2

ANNEX III.

LIST OF POSTERS

Indico ID No.	Author(s) or presenter	Member State	Poster Title	Session No.
7	J. Otoo D. Adjei E. Glover S. Inkoom K. O. Akyea-Larbi F. Otoo	Ghana	The Radiation Protection Institute Serving as a TSO for Nuclear and Radiological Education and Training	2
10	A. Athmane	Algeria	Facilitating Sustainable Education with Modern Tools in Nuclear Science and Technology, the Experience of e-Learning Cell at the Nuclear Research Centre of Algiers	3
21	S. Alnsour	Jordan	Role of Technical Cooperation in Enhancing Knowledge Management	1
28	I. Alrammah	Saudi Arabia	Proposing a Roadmap for Building Knowledge Capacity in Probabilistic Safety Assessment and Supporting Fields	7
36	S. Issaka R. Algaga M. Dotse	Ghana	Enhancing Nuclear Security Culture Through HRD: A Critical Assessment of Ghana Atomic Energy Commission	2
42	F. Braga	Brazil	Intelligence Management Cycle: An Integrated Approach to Build Organizational Knowledge and Improve Decision Making	3
44	A.B. Andam P.E. Amponsah I. Opoku-Ntim	Ghana	Nuclear Science Education and Its Role in National Development: Insights from Ghana	6
51	D.H. Nugroho B. Aji	Indonesia	Regulatory Body Knowledge Management Framework on the Licensing of Small Modular Reactors in Indonesia	4
59	S. Akyildiz	Türkiye	What the Role of Nuclear Knowledge Management (NKM) and Human Resource Development (HRD) for Research & Development (R&D) and Innovation is to Ensure a Safe and Sustainable Nuclear Programme, Including Nuclear Power, the Nuclear Fuel Cycle, Nuclear Science and Applications	4
60	R.T.P. Batubara T. Asmoro	Indonesia	Enhancing Competency and Professional Development for Nuclear Security Officers: A Case Study of the Directorate of Nuclear Facility Management, National Research and Innovation Agency of Indonesia	7
65	M. Levy	Israel	KiSure: A Professional Framework for Nuclear Expert Knowledge Retention	1
72	Y. Nam S. Yang M. Mishar H. Zhivitskaya	Republic of Korea	ANENT Online Platforms for Knowledge Sharing in Nuclear Science and Technology	5

Indico ID No.	Author(s) or presenter	Member State	Poster Title	Session No.
75	C. Anyaegbu N. Amobi U. O. Madu	Nigeria	Nuclear Knowledge Challenges and Needs Critical for the Development of Sustainable Nuclear Power Programme in Nigeria	1
80	E. Mayaka	Kenya	Advancing Nuclear Knowledge Management: Proactive Coaching and Mentoring in STEM for Enhanced Nuclear Safety and Security	6
83	R. Fornet G. Pérez H. Elias A. Machado S. Castro C. Soberat	Cuba	The Human Resources Gender Balance in the Use of Nuclear Energy in Cuba: Challenges and Opportunities	5
98	M. Almasrouhi N. Alsallum K. Alshehri	Saudi Arabia	Human Capability Development Program to Perform Effective Regulatory Functions in the Kingdom of Saudi Arabia	7
107	G. Sárdy	Hungary	Preparation of the HAEA for the Licensing of New NPP Units in Hungary	1
110	C. K. Klutse Y. Mawugbe C.Z. Dompreh A. Buckman	Ghana	Leveraging Information Technology Tools for Assessing Risk of Knowledge Loss Due to Workforce Attrition: A Case Study at the Nuclear Power Institute, Ghana Atomic Energy Commission	3
119	L. Torres F. Brollo	Argentina	The Project Radiation in Everyday Life in the RA-6 Research Reactor	4
120	H.-H. Altfeld	Germany	Introducing a Company-specific Competence Certification Scheme for Companies of the Nuclear Industry	2
133	L. Cheng W. Lin J. Qiu X. Shi	People's Republic of China	The Data-based and Intelligent Knowledge Management under the Standardized Human Resource Management System of "Hualong No.1" Nuclear Power Project	1
136	C. Carrano P. G. Mahedero J. Magits	Spain	Preserving Your Legacy – A Practical Approach to Enhancing Information-sharing, Encompassing Technology and Invaluable Tacit Expertise, across the Multigenerational Nuclear Workforce	2
142	V.K. Kuit	Malaysia	Humanizing Nuclear Science Education through Co-curricular Development	6
154	A. Albassami R. Godley	Saudi Arabia	Knowledge Transfer Methodological Approaches for Emerging Nuclear Energy Countries	2
155	M. Abaoud S. Almunahi	Saudi Arabia	Developing Behavioural Competencies for Human Resource Excellence: Integrating Behavioural Skills at NRRC	2
170	L. Vuillefroy De Silly	France	Building a Solid Skills Base for the Successful Launch of a New Nuclear Project	1

Indico ID No.	Author(s) or presenter	Member State	Poster Title	Session No.
181	A. C. Syuryavin	Indonesia	Application of Knowledge Management to Enhance National Nuclear Security Capacity Building	2
182	R. A. Da Silva Alfenas R. Baroni H. Rios C. Santos D. Silva	Brazil	Critical Nuclear Knowledge Preservation and Dissemination: The Low-cost Awarded Brazilian KM Program	1
183	N. Masngut	Malaysia	Neo Inclusivity: Leaving No One Behind	5
188	A. Timoshchenko V. Gusakova T. Savitskaya R. Sverdlov M. Tushin V. Uglov	Belarus	New Stage of Nuclear Education and Training in Belarus	5
194	L. Vuillefroy De Silly on behalf of R. Hubert	France	Industrialization of Immersive Learning at the EDF Technical Training Unit	4
204	N. Geraskin A. Kosilov E. Kulikov	Russian Federation	Knowledge Management in Nuclear Organizations: Course for University Master's Level Programmes	5
209	B. Nakyama	Uganda	Strategic Talent Acquisition for Sustainable Nuclear Workforce Development: Addressing Challenges and Embracing Opportunities. Case of Ministry of Energy and Mineral Development	1
213	V. Grance Torales F. D. Penna L. G. Lira G. A. Dalmasso F. E. Lopez	Argentina	Knowledge Loss Risk Assessment Method at CNEA – How to Achieve a New Generation Based on Knowledge	2
219	S. Litaiem	Tunisia	National Nuclear Technical Cooperation Knowledge Platforms ("National Platforms") are Established under the Global Nuclear Safety and Security Network (GNSSN)	5
228	D. Ariyanti M. Muhtadan	Indonesia	Graduate Tracer Study and Its Instrument as Collaborative Methods to Gain Insight of Curricula Input on Nuclear Vocational Education in Indonesia to Implementation NKM and Continuous Learning Improvement	6
236	M. Idris A. Rahman M. Mishar	Malaysia	Needs Assessment on Educational Sustainability in the Field of Nuclear Science and Technology: An Insight 4B	6
244	M. Karli	Türkiye	Advancing Nuclear Knowledge and Human Resources: The TENMAK Experience in Radiation Protection Training and Certification	7
246	R. Hammad	Pakistan	Reforming the Leadership Development Program (LDP) at PNRA	2

Indico ID No.	Author(s) or presenter	Member State	Poster Title	Session No.
270	N. Belmans P. Carbol M. Coeck A. Valls	Belgium	Training and Mobility in EU Projects EURAD and PREDIS to Support Knowledge Management in Radioactive Waste Management	2
274	A.A. Rahman M. Mishar	Malaysia	Enhancing Nuclear Science and Technology Secondary Education for Sustainable Development: A Comprehensive Approach and Best Practices Dissemination	6
278	M.Lamabadusuriya	Sri Lanka	Networking in Nuclear Education: Department of Nuclear Science, University of Colombo, Sri Lanka	5
283	D. Lazarevic	Serbia	Preservation of Nuclear Knowledge in Serbia	4
285	L. Lebedeva	Russian Federation	Human-centric Approach in the Nuclear Industry: from Philosophy to HR Practices	1
289	T. Toda T. Iimoto	Japan	Practical School Application of Cloud Chambers for Various Education Subjects 4B	6
290	S. Abbas	Jordan	Enhancing Secondary Education in Nuclear Science Technology: Exploring Effective Practices and Capacity-Building Strategies	6
295	Z. Tahiri	Morocco	Strategic Roles of Universities in Shaping Competence, Innovation, and Sustainability in Nuclear Education	7
305	D. Artemova N. Naumchenko	Russian Federation	Activities of Rosatom Technical Academy for Development, Implementation and Methodological Support of Critical Knowledge Preservation in Nuclear Industry	1
312	H. Touqeer	Pakistan	Role of International Training Opportunities in Human Resource Development Strategies and Nuclear Knowledge Sustainability: Transcending Boundaries	7
321	M. Coeck M.-L. Ruyssen	Belgium	A New Integrated Approach to Knowledge Management at the Belgian Nuclear Research Centre SCK CEN	4
329	N. Hussaini	Malaysia	Gamifying Radioactivity : Game-based Learning as a Pedagogical Approach for Girls in Science	4
332	R. Dolezal	Czech Republic	Some Problems of Human Resources and the Young Generation of the Nuclear Industry in Czechia (apparently nobody wants to work in nuclear anymore)	5
340	A. Kotlov	Russian Federation	Digitalization as a Natural Driver of Adaptation and Growth of Personnel Qualification to Ensure the Effectiveness of Digital Transformation of a Manufacturing Enterprise	4
342	S. Sulaiman M. Manan M. Zaini	Malaysia	Exploring Innovative Pedagogical Strategies for Malaysian Gifted and Talented Students in Nuclear Education: A Focus on Knowledge Management and Human Resource Development for Future Skilled Workforce	5
350	D. Irawan	Indonesia	Integrating Nuclear Science into Secondary Level Education Curriculum: Indonesia Experience	6

Indico ID No.	Author(s) or presenter	Member State	Poster Title	Session No.
353	H. Harun	Malaysia	Empowering Secondary Students' Interest in Nuclear Science and Technology: The Role of Counsellor Teachers	6
358	L. Ong'ayo	Kenya	Enhancing Nuclear Education and Outreach: Collaborative Initiatives among Key Kenyan Institutions	5
368	N. Kim B. Kwon	Republic of Korea	Integrating HRD and NKM Strategies in Developing Nuclear Programmes: The Korean Model for International Collaboration through Supporting Establishment of Nuclear Engineering Department in Developing Nations	5
371	S. A. Sahdan	Malaysia	Empowering Tomorrow's Minds: Challenges and Opportunities in Enriching Malaysian Education with the Integration of Nuclear Science and Technology	5
400	M. Coeck J.Stewart J. Lucey H.F.Boersma F.Draaisma J.-W. Vahlbruch F. D'Errico S.Falcon	Belgium	Policy Support Regarding Competence Building in Radiation Protection by the European Training and Education in Radiation Protection Foundation EUTERP	7
405	L. Lunan-Lyn-Cook	Jamaica	Nuclear Knowledge Management and Human Resources Development: Jamaica, Leveraging Challenges and Embracing Opportunities in an Evolving Global Landscape	4
414	G. Tikhomirov A. Kosilov	Russian Federation	The Role of Virtual Reality in Nuclear Education	4
417	A.E. Conjares A. Asuncion-Astronomo G. Poralan Jr.	Philippines	Nuclear Manpower Development in the Philippines: Experience of a Country Re-Embarking in a Nuclear Program	1
421	A. AlQubaisi	United Arab Emirates	Individual Development Plans	1
423	D. Abbasova	IAEA	A Strategical Learning as a Key Element in Nuclear Organisational Performance.	3
426	G. Balhamar	United Arab Emirates	Students Hackathon	6
431	R. Zair A. Htet T. Marfak A. Caoui	Morocco	Human Resource Development (HRD) Analysis for a Potential Nuclear Power Programme in Morocco	1
433	Z. Yu	People's Republic of China	Nuclear Science and Technology Professional Training in East China University of Technology (ECUT)	5
442	E. Fakhrurozi D. Ariyanti	Indonesia	Implementation of Tracer Study and Stakeholders or Industrial Converge to Have Recommendation of Nuclear	1

Indico ID No.	Author(s) or presenter	Member State	Poster Title	Session No.
	M. Muhtadan		Curricula Improvement on Nuclear Vocational Education in Indonesia	
449	E. Fakhrurozi A. Abimanyu K. Megasari D. Ariyanti	Indonesia	External Collaboration for Enhancing Lesson Learning of Nuclear Knowledge Via Tracer Study Method through Ten Indicators on Graduate of Polytechnic Institute of Nuclear Technology, Indonesia	5
453	L. Salvi	Argentina	Succession Planning for Critical Position Implemented at Nuclear Eleictrica SA	2
462	T. Pázmándi M. Benke G. Bruna G. Cognet B. Hatala D. Jakab C. Pesznyak A. Petruzzi K. Piliuhina	Hungary	Lessons Learned from Training and Tutoring of Experts of Nuclear Regulatory Authorities and Technical Support Organizations in non-EU Countries	1

ANNEX IV.

LIST OF SIDE EVENTS

TITLE	HOST
Atoms for Future: Talent Solutions from CNNC	China, People's Republic of
NKM Best Practices in Fusion	EUROfusion
International Mobility	France
Comprehensive Assistance to Member States in Nuclear Knowledge Management and Human Resource Development	IAEA, Department of Nuclear Energy, Nuclear Knowledge Management Section
Nuclear Education Technologies for the Net Zero Future: Rosatom Approach	State Atomic Energy Corporation "Rosatom", Russian Federation
Linking Nuclear Science and Technology into Secondary Education in Asia and Pacific	IAEA, Department of Technical Cooperation, Division of Asia and Pacific
The European Networking Effect; Enhancing Nuclear Competence Through Collaboration	European Nuclear Education Network (ENEN)
Establishment and Operation of the Nuclear Security Training and Demonstration Centre (NSTDC) in Seibersdorf followed by Technical Tour to the Nuclear Security Training and Demonstration Centre in Seibersdorf (optional)	IAEA, Department of Nuclear Security

ANNEX V.

PROSTEM CHALLENGE 2024

During the Conference the following five finalists of the IAEA proSTEM Challenge 2024 were awarded:

V-1. BEST INCLUSIVE PROJECT TO EMPOWER SPECIAL NEEDS STUDENTS

Project title: STREAM-clusive Project

Member State: Malaysia

Summary of the project:

Despite the growing importance of STEM education, underrepresented groups such as girls, children of colour and students with special needs often face challenges to engagement and participation. These challenges include limited access to advanced STEM coursers and activities, insufficient support and a lack of resources leading to low diversity in STEM fields. In 2023 we successfully organized STEM-clusive project which dedicated to engaging special needs students in nuclear science related STEM subjects. Building on this foundation, we would like to propose another value-added project, STREAM-clusive which aims to enrich the experience by incorporating two additional dimensions: Reading and Arts and broadens its reach by focusing on other underrepresented groups, including girls and children from diverse ethnic backgrounds in Asia and Pacific region. Furthermore, this project aims in making STEM more relatable and accessible by addressing diverse learning styles and cultural backgrounds challenging the perception that STEM is only for the clever or talented. The STREAM approach includes three initiatives. Developing science and mathematics through art integrating diverse reading materials into STEM and implementing project-based learning with interdisciplinary projects focused on real word problems and inclusive design. This approach aims to democratize STEM education by making it more representative of our diverse society and addressing the specific barriers faced by underrepresented groups without any academic pressure. The project plans to collaborate with educators, community organizers and technology developers to evolve and adapt the curriculum in Asia and Pacific region. Evaluation methods will be used to assess the effectiveness of STREAM and identify improvement areas. The expected outcome isa comprehensive, engaging, inclusive and practical STREAM curriculum that can engage students who might not traditionally be drawn to STEM subjects thereby enriching STEM fields with diverse perspectives and talents.

V-2. BEST OUTREACH INITIATIVE – PRESCHOOL AND PRIMARY SCHOOL STUDENTS

Project title: Genius Journals

Member State: Poland

Summary of the project:

Building Research Skills Through Interactive Storytelling Comics

The project utilizes graphic storytelling format of comics featuring scientifically accurate scenarios. After setting up an engaging and immersive narrative, a cliffhanger is presented.

The child then gets comic panels to fill out advancing the story based on their unique decisions and researched variables. As students search for answers to complete narratives, they organically develop competence in analysing problems, synthesizing information, evaluating sources and creating conclusions – key research skills. Balancing entertainment with education, the comics mesmerize through immersion in scientifically accurate worlds while secretly equipping them with abilities to explore topics systematically. The interactive experience transforms passive consumption into active knowledge creation guided by the child's unique interests. Unravelling the story becomes an exciting quest rather than an obligation.

The project leverages an interdisciplinary team spanning cognitive neuroscience for optimized science learning techniques to visual arts excellence ensuring appealing graphic execution. Neuroeducation expert Angelika M. Talaga heads learning design applying evidence-based techniques to intrinsically motivate, reinforce retention and reveal knowledge gaps. Painter and book illustrator Dominika Gołąb leads graphic conception and fine arts direction of the comics promising an unparalleled visual experience delighting and inspiring young minds. Their collaboration distils innovation across disciplines into an efficiently implementable medium that is as educational as aesthetically gripping.

This innovation encompasses producing a methodology guide for teachers and printable comic book templates enabling students to craft their own science stories. The supplemental activities can inspire classroom lessons, afterschool science clubs and summer programmes. As graphics translate universally, non-English speakers can access the learning very easily.

Through creative expression students reinforce learnings, and teachers receive insights revealing gaps in understanding. This easily implementable innovation leverages cognitive science and intrinsic motivation to cultivate research skills efficiently at low cost.

V-3. BEST STUDENT PROJECT ON STEM OUTREACH

Project title: STELLAR: Sparking Scientific Literacy, Engaging the Next Generation of STEM Leaders

Member State: Philippines

Summary of the project:

STEM fields face a critical workforce shortage, fuelled by a disengaged student population and widespread science misinformation. This project, Science and Technology Education Leaders in Learning and Research (STELLAR), aims to ignite students' passion for science and equip them with the skills to become effective communicators and future STEM leaders.

STELLAR will offer two core programmes:

(1)   Interactive Outreach Projects: Design and implement engaging STEAM activities in schools and communities, showcasing the applications of science and technology in real-world scenarios;
(2)   Research Mentorship Programme: Pair students with accomplished researchers for hands-on research experiences, nurturing critical thinking and fostering a passion for scientific inquiry.

STELLAR will leverage a multi-pronged approach:

- Peer-to-peer learning: Trained senior students will lead workshops and outreach projects, fostering a supportive and empowering learning environment;
- Collaboration with experts: Partner with scientists, educators, and science communicators to develop engaging content and activities;
- Effective technology integration: Utilize digital tools and interactive platforms to create immersive learning experiences;
- Data-driven evaluation: Continuously assess and refine programmes based on participant feedback and engagement metrics.

STELLAR builds upon existing outreach models but differentiates itself by:

- Emphasizing science communication training: Equipping students with the skills to translate complex scientific concepts into compelling narratives for diverse audiences;
- Developing research-based outreach projects: Design engaging activities that bridge the gap between theoretical knowledge and practical applications, showcasing the relevance and excitement of STEM careers;
- Building a supportive community: Foster a peer-driven learning environment where students can inspire and learn from each other, promoting active participation and leadership development.

STELLAR has the potential to become a replicable model for bridging the STEM workforce gap. By empowering students as science communicators and fostering their passion for research, STELLAR can cultivate the next generation of informed and enthusiastic STEM leaders, shaping a future where scientific literacy thrives and drives societal progress.

V-4. BEST INNOVATIVE OUTREACH INITIATIVE ON UTILIZING SOCIAL MEDIA

Project title: STEMTok – Widening STEM Outreach onto Social Media

Member State: United Kingdom

Summary of the project:

Barriers in STEM are presented to children from a young age, whether deliberately or not: boys' toys are cars and girls' toys are dolls, mathematics is hard, computer science is for nerds, physics is boring. Often these are generalisations uttered by parents, or learnt from television shows, and the impact this can have on children long term can be borne out in their own words later on: it is important to show children that science and engineering is not confined to what is taught at school and can be fun.

This project proposes a series of short form experiment videos for YouTube and TikTok that are aimed at children. Often these types of videos can be aimed at adults, instructing them on how to run these experiments with their children. Instead, this would be simple, fun science and engineering experiments presented as a cartoon for children that they can ask their parents to help with. The aim is to explain a simple experiment, such as the vinegar volcano, in short enough fashion for children to firstly understand, and second want to try with parental help. These series of cartoons can be uploaded to YouTube and TikTok and shared easily, allowing for a wide reach to potential audiences.

Children are more active than ever online, consuming a variety of media. Producing specific content to engage with children in this way allows them to see experiments, ask their parents to try, and see more about science and engineering at a young age. This interest could be further encouraged by building a coherent programme of activities to get involved in, with something new each month and engaging children with science and technology from childhood through to their university years, all in one place – a current challenge when trying to find STEM activities online.

V-5. BEST PROJECT ON AN EDUCATIONAL PLATFORM

Project title: BeSTEM

Member State(s): Russia (Zambia, India, Bosnia and Herzegovina)

Summary of the project:

BeSTEM is an online learning platform that will allow students around the world to participate remotely in specially developed courses on basic principles of STEM.

It will include two parts:

- Theoretical video-lessons;
- Materials and components for practical tasks (mini experiments).

The theoretical materials will include general modules on topics in programming, physics, mathematics, and engineering. Students will be able to watch short lessons, aimed to help them understand basic concepts, rather than memorize unnecessary information. At the end of each module, students will explain the topics using short video presentations. After the upload, the platform will give recommendations on sources that the student can use to further their knowledge. This grading system will reduce high levels of stress, as in traditional STEM education, and give them an opportunity to learn, improve their public speaking and gain confidence in the process.

Unlike existing online-learning platforms, BeSTEM will include an experimental part, where young learners can safely perform practical experiments, having regular livestreams from specialists from the BeSTEM team as guides. These live tutorials will be done in groups, with all the participants that are currently finishing the module. By performing experiments, students will not only better understand the theoretical principles, but also get closer to the real scientific work in STEM much sooner, than in traditional education.

Examples:

- Programming with microcontrollers as a practical material;
- Safe experiments, such as lava lamps (for kids);
- Making superconductors (for older students).

The platform will mainly target teenagers, and be advertised on social media, such as TikTok, Instagram, using short video forms, and done by a STEM specialist influencer. As an outcome, young people, aged from 8 to 17, will acquire knowledge efficiently and be encouraged to choose STEM as their future.

ANNEX VI.

CONFERENCE PROGRAMME COMMITTEE

Al Mansoori, S.	United Arab Emirates
Aszodi, A.	Hungary
Bales, B.	International Atomic Energy Agency
Bikkulova, G.	Russian Federation
Bonilla, N.	OECD/NEA
Chatizis, I.	International Atomic Energy Agency
Cicelj, L.	Slovenia
Cloizeaux, A.D.	International Atomic Energy Agency
Daifuku, K.	France
Danaher, T.	International Atomic Energy Agency
Delgado, J.L.	Spain
Delmotte, A.	United States of America
Di Trapani, A.	Italy
Dias Acar, M.E.	Brazil
Diéguez Porras, P.	International Atomic Energy Agency
Gloeckler, O.	International Atomic Energy Agency
Hassan, M.	Egypt
Huang, W.	International Atomic Energy Agency
Iunikova, A.	International Atomic Energy Agency
Jevremovic, T.	International Atomic Energy Agency
Michal, V.	International Atomic Energy Agency
Moracho, M.	International Atomic Energy Agency
Nakano, Y.	Japan
Nam, Y.	South Korea
Neuhold, M.	International Atomic Energy Agency
Pavel, G.	European Nuclear Education Network (ENEN)
Reysset, T.	International Atomic Energy Agency
Sari, S.	Türkiye
Schoevaerts, D.	International Atomic Energy Agency
Shumski, V.	International Atomic Energy Agency
Sowagi, J.	Canada
Subki, H.	International Atomic Energy Agency
Wang, Y.	China, People's Republic of
Whan, S.	International Atomic Energy Agency
Young, A.	International Atomic Energy Agency

 **IAEA**
International Atomic Energy Agency

Feedback on IAEA publications may be given via the on-line form available at:
www.iaea.org/publications/feedback
The form may also be used to report safety issues or submit environmental queries concerning IAEA publications.

Alternatively, contact IAEA Publishing directly:

Publishing Section, Dissemination Unit
International Atomic Energy Agency
Vienna International Centre, PO Box 100, 1400 Vienna, Austria
Telephone: +43 1 2600 22529 or 22530
Email: sales.publications@iaea.org
www.iaea.org/publications

ORDERING LOCALLY

To purchase priced IAEA publications, please contact either your preferred local supplier
or the IAEA's lead distributor:

Mare Nostrum Group

39 East Parade
Harrogate
North Yorkshire
HG1 5LQ
United Kingdom
Email: enquiries@mare-nostrum.co.uk
www.mngbookshop.co.uk

Trade Orders and Enquiries:

Telephone: +44 1243 843 291
Email: trade@wiley.com

Individual Orders and Enquiries:

Email: mng.csd@wiley.com

Priced and unpriced IAEA publications may also be ordered directly from the IAEA by contacting
IAEA Publishing. The recipient is responsible for shipping costs and any customs and duties.

Printed and bound by CPI Group (UK) Ltd, Croydon, CR0 4YY

06/07/2026

02160601-0001